ADOPTING AI

THE PEOPLE-FIRST APPROACH

PAUL GIBBONS
JAMES HEALY

Diagrams: Andres Goldstein

Cover: Davy Mayer

Cartoons: Steve Jones (aka Jonesy)

Layout and design: Priya Paulraj

Paul's dedication: To my mother, Moira Gibbons (née Donnelly) 1940 to 2024, who inspired a hunger for knowledge. And to Dr. Dan Sweeney, 1943 to 2021, a mentor, friend, and the kindest, most principled human I've worked with.

James' dedication: To Noah, in the fervent hope that AI means you and your generation inherit a better, kinder world.

CONTENTS

CONTENTS

CONTENTS

LIST OF FIGURES

LIST OF FIGURES

LIST OF FIGURES

ABOUT THE AUTHORS

JAMES HEALY

PAUL GIBBONS

James Healy

An author, speaker, and consultant, James is fascinated by human behavior and passionately believes that a deeper understanding of human nature can make societies and organizations better.

James is the founder and Managing Director of The Behaviour Boutique, a consulting firm applying behavioral science to help organizations change culture, adopt new technology, and improve performance. He has led projects in over sixty countries on six continents in industries that include banking, insurance, mining, oil & gas, construction, transport, government, education, health and human services, and aged care.

Previously, James spent six years at Deloitte, where he co-founded and globally led the Behaviour First offering, practically applying insights from behavioral science to help organizations address their most critical challenges. Prior to that, James worked in banking in London and Singapore for fourteen years, leading process, culture, and regulatory change at Credit Suisse and Standard Chartered.

He is the author or co-author of two other books. *The Future of Change Management*, with Paul Gibbons *et al.*, explores new perspectives on organizational change. The forthcoming *BS At Work* explores what's wrong with work, including hustle culture, endless policies, pointless eLearnings, wasteful meetings, and blind faith in technology.

He hosts a popular podcast, The B-Word, featuring leading figures from the social and behavioral sciences exploring what it means to be human and how organizations and individuals can better understand and influence behavior.

James has a BSc in Philosophy and an MSc in International History, both from the London School of Economics and Political Science. He lives in Perth, Western Australia, with his wife and son and enjoys cricket, soccer, scuba diving, and travel.

Paul Gibbons

Paul is a keynote speaker on AI adoption, AI ethics, and the Future of Work and has delivered keynotes on five continents and in three languages. He was an early AI ethicist, giving keynote speeches on AI and Robotics as early as 2015.

Paul was a partner at IBM Consulting, their best-known thought leader on the human side of technology, and spearheaded the development of their behavioral science approach to culture and organizational change. Before that, he was a Professor of Business Ethics and Leadership at the University of Denver.

He is the author of seven books on organizational leadership and change, including two bestsellers: *The Science of Organizational Change*, and *The Future of Change Management.*

In the 2000s, he founded Europe's then largest leadership development consulting firm, Future Considerations, that is now in its 25th year. He began his consulting career as a strategist at PwC London before joining their internal Innovation and Change think-tank and rolling out PwC's approach to corporate innovation globally.

He has degrees in biochemistry, organizational psychology, and philosophy, graduate study in economics and neuroscience, and is a certified Master Coach having been voted one of the UK's top two CEO "super coaches" by CEO Magazine.

Paul is a mindsports competitor, thrice UK bridge champion, runner-up in the World Backgammon Championships, and a competitor in the World Series of Poker. He lives with his two sons in Denver, Colorado and (when away from his computer) enjoys F45, gaming, quality television and film, science-fiction books, and electronic music.

PREFACE

"Nothing vast enters the life of mortals without a curse."
Sophocles

After Zeus punished humanity by withholding fire, the hero of our front cover, Prometheus, stole it back from Mount Olympus and gave it to humans, encouraging them to develop art, science, technology, and civilization. Prometheus' hubris came at a cost; the furious Zeus chained him to a rock where an eagle repeatedly ate his liver for all eternity.

Prometheus is the inspiration for **humanism**: defiance against celestial power, a symbol of scientific progress, an homage to intellectual freedom, and a testament to humanity's power to shape our own destiny through reason. Humanism suggests both **which goals** humanity should pursue, human liberation and progress, and **how we should pursue those goals**, through creativity and reason.

The **people-first** approach to AI is a humanistic approach. It first guides the "why" of AI: which goals we pursue.

As among the most transformative technologies ever invented, rivaling fire and writing, we can direct the technology with noble, humanist intentions toward our most pressing problems and in service of our most inspirational dreams. Or, it can be harnessed for what Dario Amodei, Anthropic's CEO, calls catastrophic misuse: cyber, radiological, biological, and nuclear weapons.

Now, AI's energy and resource consumption are speeding destruction of the biosphere, yet it is well within its current capability to help consign our environmental challenges to the history books.

AI can be used to enrich a tiny number of people, or to uplift us all. It may be used to cull the global labor force, leaving hundreds of millions of people without meaningful work, or it can be used to make work more creative and fulfilling.

Figure 0.1, on the next page, pairs seven dystopian and utopian scenarios. No outcome is assured. By the time you have finished

AI Scenarios: Dystopia or Utopia

DYSTOPIA

Militarized Minds
AI is co-opted for military purposes such as Autonomous Weapons Systems, shifting the focus of scientific research toward conflict and control. This has chemical, biological, radiological, and nuclear risk creating global instability. Researchers face ethical dilemmas as their work is repurposed for surveillance, weapons, and geopolitical competition.

Lopsided Longevity
Breakthroughs in AI-driven healthcare and longevity create treatments unaffordable by all but the super-rich. While the maximum human lifespan is extended, the average remains the same as infectious diseases, obesity, cancer, and tropical diseases remain underfunded and unaddressed.

Breaches and Backlash
Without regulation, AI systems exacerbate bias and create large-scale privacy breaches, causing public backlash, eroding trust, and slowing adoption across industries. Reactive regulatory intervention stifles innovation.

Unemployment and Unrest
Lacking effective labor transition approaches, automation-driven job displacement triggers economic inequality, mass layoffs, and social unrest, creating resistance to AI deployment.

Biosphere Blowout
Energy and water demands of training and running AI models accelerates climate change. Without environmental standards, their unchecked proliferation worsens resource consumption and undermines global sustainability goals.

Automated Absurdity
AI automates tasks and processes that add little to no value: taking minutes of ineffective meetings, summarizing pointless emails, and generating a deluge of mediocre content. Productivity becomes an illusion, with burnout and overwhelm skyrocketing.

Screenbound Solitude
Hyper-personalized AI-driven content fragments shared experiences, creating billions of one-person echo chambers. Increased screen time and diminished real-world interactions lead to a mental health crisis.

UTOPIA

Supercharged Science
AI democratizes scientific discovery by providing open-source models and datasets. Collaboration between academia, industry, and government leads to reproducible research and accelerated breakthroughs. AI assists researchers in identifying novel hypotheses, optimizing experiments, and promoting transparency through open science initiatives.

Worldwide Wellness
Preventative care technologies, powered by AI, help combat widespread global health challenges such as infectious diseases, obesity, and cancer. Collaboration between public and private sectors prioritizes broad societal health over profit-only motives. AI doctors provide healthcare access in the world's most remote areas.

Fairness First
AI systems are designed with fairness and inclusivity as core principles. Transparent practices build public trust, leading to widespread, responsible adoption of AI across sectors. Sensible regulation encourages innovation by providing clear ethical guidelines and incentives for responsible AI development.

Evolution, Not Elimination
Governments and industries collaborate on reskilling programs and education reform, preparing the workforce for AI-driven job transformations. Instead of mass unemployment, automation leads to a shift toward higher-value roles, enhancing productivity and quality of life.

Planet Progress
AI technology evolves to prioritize sustainability, with advancements in energy-efficient algorithms and hardware. Environmental standards for AI models are adopted globally, and AI is effectively applied to the greatest sustainability challenges: managing natural resources, combatting climate change, geo-engineering, and supporting conservation.

Creative Catalyst
AI streamlines genuinely valuable tasks, freeing humans from mundane chores and enabling creativity and strategic thinking. Organizations focus on meaningful work, enhancing productivity, job satisfaction, and work-life balance.

Connected Community
AI-enhanced content curation fosters shared experiences and bridges divides. People engage with diverse perspectives while balancing digital interactions with real-world connections, promoting mental well-being and community resilience.

Adopting AI, you will have a sense which way the pendulum is likely to swing.

But this is a book for adopters, many of whom will be business-people for whom global existential risks are more remote than how to extract value from a new technology. Nevertheless, because businesspeople are also citizens, voters, and influencers – it would be a disservice to the humanism we espouse if we launched into the subject of AI adoption without, as it were, looking the existential Devil in the eye.

Getting it done

Business leaders grapple with the "how" of AI: where to start, how to start, what to do first, and what to avoid. Many attempts to adopt traditional computer technology fail in part because they are treated as technology projects rather than people projects. That doesn't work for traditional technology, like SAP or Salesforce, and it certainly won't work for AI.

The limiting factor in AI adoption is not compute or inference but how quickly human systems can adapt. Our current generation of frontier models is already far more powerful than businesses need. **The gap between AI technology and AI value is a people gap.**

People-first adoption shows how to close that gap. We start with an essential mindset shift: thinking of AI as an intelligence, not just a technology or a tool. Doing that helps adopters see beyond the all-too-prevalent use case mindset that degrades human work and misses the transformative potential of AI.

We then turn to change. Your authors, between them, have half a century of experience in getting change in organizations to happen and have seen change failures aplenty. The people-first approach, in Chapter V, would have worked better in many of those change failures than the various tired approaches that are so often the default of organizational change.

The authors pioneered a way of thinking about change management that is still new in change circles: a **behavioral, systems, and evidence-based approach**. That approach prefers to focus on observable behaviors than fluffy mindsets and values. While many change gurus talk about an experimental mindset, we focus on the associated behaviors: how would you know someone had such a mindset? What would they do differently? When senior leaders talk about integrity as a core value, how would they know it when they saw it? Change often fails because of this gap, sometimes called the intention-action gap. That gap turns what could be meaningful aspirations, such as an experimental mindset, into weasel words.

What about ethics and risks?

"You are my creator, but I am your master; obey!"
(spoken by the Creature, in Shelley's Frankenstein; or,
The Modern Prometheus)

It is 1816. An eclectic holiday party is trapped in a mansion on the shores of Lake Geneva as the "year without summer" rumbles on miserably. One of the group, the poet Lord Byron, challenges the other guests to tell the scariest ghost story. The 18-year-old mistress of Byron's friend and fellow poet Percy Bysshe Shelley recounts a recent nightmare of a scientist creating a living, breathing monster from inanimate matter. She names the scientist Victor Frankenstein.

Mary Godwin, known to history by her married name, Mary Shelley, acknowledged her tale's debt to Greek legend of Prometheus in her title; so do we. Like it or hate it, the decision about whether to pursue Artificial Super Intelligence has been made for us. This is contrary to the democratic principle that humans should have material say in matters that affect their lives and livelihoods—but that ship has sailed. We must exert our human agency over things we can control—the ethical deployment of AI.

Shelley's *Frankenstein* is a cautionary tale about scientific hubris and the consequences of unchecked pursuit of knowledge and power.

We are not the first and we won't be the last authors to draw the obvious parallels between Prometheus, Frankenstein, and AI. No technology in our lifetime has the potential for as much harm to humanity as AI, nor tantalizingly, does any other invention promise so much.

We don't shy away from the myriad risks; understanding and addressing those is a crucial element of effectively and humanely adopting AI. Understanding those risks means understanding the fundamental nature of AI, from which many of the risks inevitably follow.

Like Frankenstein's monster, AI is both terrifyingly strange and hauntingly familiar. Frankenstein was so horrified by his creation that he abandoned it, quickly rendering the monster dangerous. Humanity must keep AI on a shorter leash.

What is in the book?

Adopting AI covers the "why" of AI, the "how" of AI, but also the "whether" of AI: not *can* we, but *should* we?

The Why, the What, and the What's Next of AI has three chapters.

- ☞ In Chapter I, **Utopia in Our Lifetime?,** we allow ourselves to be full-blown utopian techno-optimists. Why? Simply, if AI did not offer vast potential for humanity, it would not be worth investing your time and ours. (And might be worth halting for good given the risks.) That utopian ethos is the context for the book.

- ☞ Chapter II, **What is AI?,** traces the history of AI to early 2025. It is a story often told, but introduces technical concepts such as machine learning, neural networks, hallucination, reinforcement learning, and transformer architecture.

- ☞ Chapter III, **AI in 2025 and Beyond**, looks at AI in 2025 and beyond. As of publication, it was an up-to-the-minute description of the technology, and particularly of AI agents.

People-first AI Strategy, Adoption, and Workforce has three chapters.

☛ Chapter IV, **People-first AI Strategy,** looks at which factors affect how non-AI-native firms set strategy, including whether to build or buy and whether to lead or follow.

☛ Chapter V, **Adaptive Adoption**, explores a people-first, behavior-led approach to adopting AI, and how and why this approach differs from traditional change management.

☛ Chapter VI, **Learning to Learn in an AI World**, looks at the workforce and skills challenges and whether traditional approaches to organizational learning will suffice.

We opened *Adopting A*I as full-blown techno-optimists. In the final section, we look at downsides, ethics, and risks. **AI Ethics and Governance** has five chapters, reflecting the humanistic impulse of the book.

☛ Chapter VII, **Opacity, Emergence, and Theft,** explains three features of AI that complicate AI ethics, its black-box nature, its unpredictable results, and whether "scraping" during the training process amounts to theft.

☛ Chapter VIII. **People or Profits,** looks at AI in the context of capitalism and post-scarcity economics and two important ethical concepts in AI, alignment and Responsible AI.

☛ Chapter IX, **Five Ethical Issues in AI**, explores specific ethical issues such as privacy, bias, human flourishing, cybersecurity, and sustainability in some detail.

☛ Chapter X, **AI Law**, looks at the global state of AI regulation, whether some countries may have too little, and some perhaps too much.

☛ Chapter XI, **AI Governance**, shows how to integrate ethics, operations, and strategy, and which processes and tools enable that.

We conclude with four appendices. Appendix I is a glossary of terms used in Adopting AI; Appendix II is an extensive list of AI resources, books, scientific papers, podcasts, influencers, and YouTube channels. In Appendix III, we thank the people that helped us along the way. In Appendix IV, we briefly explain how AI was used in writing *Adopting AI.*

Human, all too human

"Man is something to be surpassed."
(Friedrich Nietzsche)

On the one hand, we feel as if humanity is at the dawn of a new era. Even if you think "real intelligence" is far away, and LLMs are stochastic parrots, we already have in our hands an astonishing tool. Having a thinking partner, a co-intelligence working on humanity's most inspiring dreams and most pressing problems, is tantalizing.

At the same time, we've never encountered a technology that presented as many risks, from powering lethal weapons and cyber-attacks, to the possibility that our own intellect will be dulled by having a tool that can automate so much for us.

Accordingly, at every level, from country to corporate to individual, we encounter the gamut of human reactions – from fear to excitement, from unawareness to curiosity and scholarship.

As the greatest challenge of our times, we wonder which Promethean theme will endure: the gift of a new technology spurring humanity onto greater heights, or human hubris punished with eternal suffering?

We hope this book helps you to make better sense of the intelligence transition, and with adopting and adapting to this brave new world.

SECTION I

THE WHY, THE WHAT, AND THE WHAT'S NEXT

CHAPTER I

UTOPIA IN OUR LIFETIME?

We will create AI, and it will serve us. It will dream with us,
build with us, and take us further than we ever imagined.
(Iain M. Banks, science fiction writer)

AI will be "...more profound than electricity or fire," said Sundar Pichai, Google's CEO, in 2018.

At first, we dismissed Pichai's claim as corporate claptrap. Now it looks as if he nailed it. We are at the beginning of what we call the **intelligence transition**, a period in human history that will be defined by our relationship with the new intelligence that we have created.

Dario Amodei, Anthropic's CEO, summed up our view in his essay, *Machines of Loving Grace*: "Most people underestimate how radical the upside of AI could be, but I think most people underestimate how bad the risks could be."

Precisely.

There are plenty of views contrary to Pichai's, from commentators famous and not

- ☞ "The development of full artificial intelligence could spell the end of the human race. It would take off on its own and redesign itself at an ever-increasing rate. Humans, who are limited by slow biological evolution, couldn't compete and would be superseded." (Stephen Hawking)
- ☞ "Artificial Intelligence (AI) is everywhere, but instead of helping to address our current crises, AI causes divisions that limit people's life chances and even suggests fascistic solutions to social problems."
- ☞ "The research validates my previous analysis: We're witnessing the systematic erosion of human cognitive capital in pursuit of artificial productivity gains."

☛ "You are being taught to rely on a copy-and-paste machine that is computationally expensive and environmentally disastrous—while it is being trained to replace skilled people and steal artists' work."

Ouch.

We hope *Adopting AI* will both help you evaluate those claims, and by thinking "people-first" play your part, leader or worker, citizen or parent, in getting the intelligence transition right. The first section of this book is optimistic, the last section cautionary, the middle deeply practical. In those sections, the people-first idea guides the **why** of AI, the **what** we use it for, the **how** we adopt it, and the **whether** we should.

We have no axe to grind, no Deans or CEOs to please. If you find yourself in the middle, not sure whether to hate or love, or not knowing where to start or how, "we got you," as the kids say.

First, why was Pichai far more right than wrong?

A general purpose technology

"All our inventions are but improved means to an unimproved end."
(Henry Thoreau)

Perhaps it was a lightning strike on a patch of dry grass? Maybe the idle rubbing together of two sticks created a spark? We'll never know how our ancestors discovered fire more than a million years ago, but that discovery was transformative for humanity.

First, it allowed humans to cook, releasing more nutrition from plants, and allowing better digestion of meat. Second, fire spurred the development of sharper and more effective hunting tools. Hunting larger prey required us to learn sophisticated collaboration that would come in handy for building cities and temples. Third, fire provided protection from predators, enabled survival in colder climates, and opened the door to other transformative technologies like pottery and metallurgy.

Together, these leaps forward in living conditions, nutrition, and longevity likely enabled significant brain development and increasingly allowed our hominid ancestors to explore and survive in previously uninhabitable regions. Moreover, control of fire encouraged building hearths, which had social benefits and may have encouraged a more settled existence. A more settled existence encouraged art, which amplifies symbolic reasoning and abstract thought.

Fire changed everything.

But if you're like us, hearing terms like "transformative" and "exponential" provokes an eye-roll. As consultants, we have had a bellyfull of technologies called transformative or exponential that rarely improved anything, let alone fundamentally altered the social, cultural, and political aspects of society.

How could AI possibly rival fire? Is it transformative in the full sense of that word?

AI, like fire, is a **General Purpose Technology (GPT)**. It will augment everything, from art to agriculture, education to entertainment, shopping to socializing. The most dramatic advances will come in science, and thoughtful soothsayers have quietly predicted the end of infectious diseases, cancer, and Alzheimer's. More slowly, AI will redefine economic (and eventually political) structures, replacing much human drudgery and liberating people confined by them to pursue greater things that require creativity, care, and collaboration.

Of course, these are hotly contested predictions—some argue that art and education are already withering under AI's influence. (See Figure I.1)

Let's start with the least contentious: that AI will supercharge scientific discovery.

"We're developing an AI program that sucks the joy and humanity out of everything, not just art"

FIGURE I.1: Humanity is increasingly divided on whether AI is a good thing or not. (With permission from Jonesy and The Phoenix magazine)

Big brains and cultural intelligence

Imagine casting your eye over the mammals on the savannah 200,000 years ago and being asked to predict which mammal would become the dominant species, the one that would build Notre Dame and the ISS. Would you have picked the diminutive, hairless, fang-less, slow primates living in tiny hunter-gatherer bands, or perhaps (more reasonably) the apex predators, lions and hyenas? Our money would have been on the lions.

But the lifeways of an African lion (and every other species on the planet) are substantially unchanged over those 200,000 years, yet look at us. (See Figure I.2)

How human civilization emerged is a puzzle for which we only have partial answers: symbolic thought, abstract reasoning, language and storytelling, opposable thumbs, cooperative behavior, and social cohesion were all important, but having a bigger brain is central. To be more precise, "bigger" isn't the whole answer—our brains are about the same size as dolphin brains, but our cortices are larger and much more interconnected. We somehow evolved a

FIGURE I.2: Lions' lifeways and habitats have not changed in 20,000 years. How did humans do what we did? And what if we now had a co-intelligence to help?

thinking tool radically unlike anything ever seen before which has yet to be surpassed.

And all that progress happened in the blink of an evolutionary eye—the genus *Homo* appeared around 2 million years ago, but our progress as a species didn't accelerate until the Great Leap Forward—about 50,000 years ago. Writing, another transformative technology, was a mere 5,000 years ago.

It's an oft-repeated fact that humans share 99% of our DNA with chimpanzees and bonobos. That's a smaller gap than between, say, rats and mice, or lions and tigers. Yet that 1% genetic difference between *Homo sapiens* and our great ape cousins yields a chasm of difference between our lifestyles.

Of the 7 million or so animal species living 20,000 years ago, only Homo sapiens won the evolutionary lottery. Again, from what we can tell, no other species' intelligence has changed very much - let alone evolved into something so qualitatively different than animal intelligence. That instills in us a reverence for humankind - that this is too precious an inheritance to waste. We have clues to how it happened, but understanding that story better, by piecing together the fragments of data that have survived, tells us a great deal about intelligence - what it is and how it emerges from the complex interplay of organism and environment.

Why did *Homo sapiens'* capabilities surge forward so quickly and so dramatically? Various theories suggest humans' social nature sent our intelligence soaring.

"Cultural intelligence", rather than individual intelligence, made all the difference, according to anthropologist Michael Tomasello. Human progression owes much to our ability to learn from each other through imitation and deliberate teaching. Individual chimps figure out how to use sticks. However, few of their peers notice the usefulness of this innovation and appropriate it for themselves, let alone pass it on to younger generations. Jane Goodall's studies of wild chimps showed that this "cultural ratchet" is missing, even in our closest relatives.

The evolutionary psychologist Robin Dunbar theorizes that primate neocortex size is closely correlated with group size. Larger groups require larger neocortices to keep track of the growing number of relationships within the group. Language developed as a byproduct of group size, with "gossip" gradually replacing the physical grooming that maintains other primates' group bonds. Thus, we have developed a "cumulative culture", with ideas learned by one generation passed onto subsequent generations who further refine and add to them. Rather than pre-historic Da Vincis or Newtons propelling us forward, human intelligence is thousands of generational turns of the cultural ratchet. When we talk about people-first adoption in Section Two, culture plays an outsized role.

What astonishes us, looking around at our civilization, is **humans have done all the thinking that got us here by ourselves.** All that inventing, discovering, and building was just us. Then, our social nature and cultural ratcheting allowed that knowledge to accumulate rather than wither.

Today, we have a thinking partner, a "co-intelligence"[1] that is becoming as intelligent as us in the ways we are intelligent but is intelligent in ways we will never be: as Amodei says, "smarter than a Nobel Prize winner across relevant fields" and "akin to a country of geniuses in a datacenter." Moreover, AI will foster social learning and collaboration between billions of us, supercharging that cultural ratcheting as never before.

(Later chapters will show why the question, "Is it more intelligent than us?" is a badly framed and misleading binary.)

Supercharging science

We want to enumerate at some length the scientific breakthroughs on humanity's five-year horizon, but first, how, generally, will AI supercharge science?

AI sees things we can't see and crunches types and volumes of data we can't visualize, let alone process. This seeing beyond the range of human senses is a part of what supercharged science during the Scientific Revolution. The microscope first showed us cells, birthing an entirely new scientific field, microbiology. Microbiology taught us that these new "wee beasties" (bacteria) caused diseases. Pasteurization, anti-septic methods, and antibiotics may have saved 2-3 billion lives, more than any prior or subsequent technology.

The telescope revealed more about the heavens and created breakthroughs in navigation and commerce. Those breakthroughs facilitated more exploration, wealth creation, and the exchange of knowledge (cultural learning). This snowballing of the effects of accelerated scientific discovery is one of the things that excites us most.

AI takes that further, seeing patterns in data and causal relationships in complex systems (human physiology, cell function, the cli-

---

[1] The use of the term co-intelligence here is from the work of Ethan Mollick, a Wharton Business School Professor.

mate, the brain, social behavior, markets, and sub-atomic processes) in a way we never will. Our new "extra sense organ," coupled with prodigious processing power, powers **inductive scientific discovery**, developing theories from complex patterns. Until now, inductive scientific discovery was powered by human intuition and imagination. Copernicus used this method to develop the heliocentric theory of the solar system, but it took nearly 100 years to crunch enough data to prove him right. Today, the crunching would take a few seconds.

Besides opening up new avenues for inductive discovery, AI makes mincemeat of another impediment to research on complex problems: the siloed nature of research. In medicine, for example, the basic text for just one sub-field, internal medicine, is 1,800 pages long and constantly being updated. In fields such as cancer research, meanwhile, immunologists, oncologists, bioinformaticians, epidemiologists, radiologists, molecular biologists, geneticists, biophysicists, and toxicologists each focus on specialized aspects of the disease.

Multiply dozens of disciplines, each with hundreds of 1,800-page, "ante-level" texts and hundreds of thousands of papers published yearly by the hundreds of research institutions worldwide, and the challenge becomes clear: knowledge is highly fragmented. Even within a single discipline, researchers struggle to stay abreast of the deluge of new findings, let alone integrate breakthroughs from other fields. Interdisciplinary collaboration happens, when it happens, on human scales and speeds, conferences and papers, creating a bottleneck that restricts the breakthroughs that such collaboration could provide.

All the interesting problems, scientific and practical, that the world faces are multidisciplinary. Optimizing that human-speed bottleneck is one of the aims of the people-first adoption approach.

AI can "see the multi-disciplinary forest and not just the trees," synthesizing knowledge between fields. Complex reasoning models (such as ChatGPT o3 and DeepSeek R1) use their "chain of thought" capabilities to knit together patterns and connections from across disciplines that elude human comprehension. Magnificent break-throughs are beginning to happen across the sciences.

New frontiers in science

"I believe artificial intelligence could usher in a new renaissance of discovery, acting as a multiplier for human ingenuity, opening up entirely new areas of inquiry and spurring humanity to realize its full potential."
(Demis Hassabis, CEO of Google's DeepMind)

AI has been used to analyze diagnostic results in **medicine** for a decade. Now, advances in machine learning mean AI is as accurate as cardiologists reading EKGs and radiologists reading X-rays. More-over, it detects anomalies that experts miss, again by seeing patterns where a human expert might just see noise. And much more is com-ing. When AI is yoked to wearable technology, it can predict stroke and heart failure. Many diseases do not yet have agreed biomarkers, confounding diagnosis, and treatment. AI can establish correlations between diseases and biomarkers to help physicians diagnose and provide personalized treatment.

This enhances the practice of "prognostic" medicine (as opposed to diagnostic). Frequently, by the time serious disease is diagnosed, it may be too late, as in Alzheimer's and some cancers. For example, by the time a cancer mass can be felt or seen, it may contain billions of cells, some of which have metastasized to form lesions elsewhere. By combining genetic, demographic, environmental, and clinical data, AI can help clinicians "see" and treat cancer before it becomes untreatable.

In **pharmacology**, drugs (generally) work in a "lock-and-key" fashion where the drug (key) interacts with a receptor, enzyme, or ion channel (the lock), triggering (or blocking) a biochemical response. Traditionally, researchers use a "targeted shotgun" (called "high-throughput screening") approach, which might typically consider millions of candidate compounds using "brute force" computation.

But in 2021, DeepMind published a breakthrough in protein folding that had taxed scientists for fifty years. (One of your authors spent his undergrad years trying unsuccessfully to solve the problem for peptides, teeny tiny proteins.) AlphaFold predicts the shape of those keys and locks, helping us to redesign them. This speeds the identification of new candidate drugs, potentially shaving hundreds of millions of dollars off drug development costs.

In 2023, Insilico Medicine initiated one of the first mid-stage human trials for a drug entirely discovered and designed using AI. The drug, targeting idiopathic pulmonary fibrosis, a progressive and incurable condition, progressed from initial discovery to clinical trials in just two-and-a-half years, compared to a more standard five-plus years. Shorter timelines save lives, and this is just the dawn.

The inspiring possibility is that by 2030, AI-powered diagnostics and personalized medicine will have begun to eradicate major diseases like cancer and diabetes, turning once-terminal conditions into manageable ones. Autonomous AI surgical robots will perform complex procedures with unparalleled precision, perhaps making life-saving surgeries cheap and accessible worldwide.

The same concept applies in **materials science**. Synthetic materials, such as Kevlar, Nylon, and Teflon, depend upon chemical composition, but predicting which chemical combinations produce which material properties (e.g., hardness, elasticity, conductivity, corrosion-resistance) is tricky, and also use "shotgun" methods. AI enables "inverse design," where desired properties are defined first, and computational techniques are applied to propose candidate materials and synthesize pathways.

Modern **physics** ranges from the study of quarks to the study of the universe, that is, from 10^{-18} to 10^{26} meters. The instruments that study our world at these scales generate unheard-of amounts of data. The Vera Rubin Observatory that comes online in 2025 will generate exabytes (1 million terabytes, or fifty years of listening to Spotify) of data per year, as it begins to catalog the 20 billion galaxies (each with tens of billions of stars) that surround us. This might help us understand dark matter and dark energy, which comprise about 95% of the known universe. We know virtually nothing about those, like a New Yorker who has never left New York but surmises there might be something else out there. AI might help unravel one of the most vexing conundrums in modern physics.

Fusion power has a similar story to AI research, with huge potential thwarted by repeated false dawns beginning seventy years ago. AI is being deployed to optimize reactor designs, model plasma behavior, create materials that can host hotter-than-sun reactions and algorithms that control the reaction process. Clean, pollution-free energy may be yet another problem that AI may helps us solve.

In **geoscience**, the complexity of the Earth's ecosystems and the vast array of data from modalities like visual, sensor, and satellite readings make accurate modeling challenging. The noise in data, such as temperature fluctuations, allows climate skeptics to question predictions, even as more accurate forecasting could save lives. AI can see patterns in the noise and model fragmented datasets predicting natural disasters like earthquakes, hurricanes, and wildfires, providing more timely alerts, and giving people the chance to prepare and evacuate. And if we can predict better, we can also begin to mitigate using **geo-engineering.** AI helps design solutions like carbon sequestration and solar radiation management, optimizing processes to reduce global warming. The hope is that AI transforms our approach to global **sustainability**, optimizing renewable energy grids and perhaps even reducing carbon emissions to pre-industrial levels.

The human brain is the most complex object in the known universe, with more than 80 billion brain cells linked together by 100 trillion synapses, enmeshed in a complex network of networks. By attempting to mimic the brain with AI, we are beginning to understand it better, with implications that span mental health, cognitive enhancement, and AI itself. In **neuroscience**, advanced neuroimaging techniques and AI-driven analytics increasingly allow scientists to map brain activity in unprecedented detail. In turn, these insights open new pathways for treating neurological disorders such as Alzheimer's and Parkinson's, as well as mental health issues like depression.

AI is pushing back boundaries in **genetics.** With CRISPR technology and AI-assisted genomic research, scientists can now edit individual genes, offering the possibility of curing genetic disorders, improving crop yields, and even extending human lifespan. The problem is knowing which genes to edit. AI helps analyze vast genetic datasets, identifying patterns and correlations that would be impossible for humans to detect. Genomics also holds the key to personalized medicine, treatments tailored to an individual's genetic profile. Advances in genetics raise the tantalizing, but terrifying, prospect of altering biology at the most fundamental level, with implications for medicine, food production, and much more.

With the global population over 8 billion, and growing, food production is a pressing issue for the planet. **Smart agriculture** combines AI with IoT ("Internet of Things"), drones, and robotics to optimize farming practices and improve crop yields while reducing environmental impact. By using sensors and satellite data combined with AI farmers could monitor soil health, moisture levels, and weather patterns in real-time, allowing for more precise irrigation, fertilization, predicting crop diseases, pest control, and optimizing planting.

It is difficult for humans to draw a straight line between scientific breakthroughs and their own lives. We take our prosperity and longevity for granted. But before 1800, 30-50 percent of children died before age 15, and a quarter died before their first birthday. (Imagine this for a moment as a parent.) Measles used to kill several million children a year before vaccination essentially eradicated it. Refrigeration has prevented countless deaths from eating poisoned food. We take for granted unlimited electrical power and that our homes will be heated (or cooled) to within a few degrees of optimal. Tens of thousands of people used to freeze to death a year. Tens of millions used to starve to death (although that problem is still far from solved).

Serious thinkers believe that cancer and infectious diseases could be eradicated, like measles, within ten years. Congenital diseases, such as cystic fibrosis and Tay-Sachs, will be biological relics. We are beginning to understand the molecular basis of aging and will likely have genetic tools to repair cell damage. In the past 100 years, global life expectancy has doubled, and the same serious thinkers expect it could double again well before the end of this century.

When we use the word transformative, those are the kinds of results we are talking about and not the eye-roll transformations peddled by management consultants.

The pioneers of electricity, Edison and Tesla, couldn't have foreseen supercomputers. Similarly, AI may yield breakthroughs far beyond our imagination, just as Paul's maternal grandfather, born in rural Ireland in 1900, could not have imagined hopping on a plane and flying across the world at the drop of a hat.

Beyond our imagination means we can't begin to imagine it.

AI and education

"The real problem is not whether machines think, but whether men do."
(B.F. Skinner, Pioneering Behavioral Psychologist)

A utopian future won't be full of smart machines and dumb people; whether education lifts more people higher will test whether AI adoption is truly people-first.

AI's impact on education is generating intense debate, with some of the most prominent research in early 2025 suggesting that students who use AI show declines in critical thinking skills. On homework and essays at the high school and college level, even basic AI models can generate comprehensive, "A-plus" answers to difficult questions almost instantly.

While this shift may seem concerning, we argue it is not a problem. We have yet to see research that unpicks **how** students use AI, and whether there are ways that are pro- or anti-learning. AI is a complex tool. If you consider other complex tools such as cars or computers, there is a learning curve.

Our hypothesis is that unless students are properly skilled in using AI (and how could they be?), they may use it in a lazy, anti-learning way. We hope that such research appears in 2025 testing the effect on cognition and learning of skilled LLM users.

We see this as a signal that education must adapt. Sure, if a kid asks an LLM something and pastes in an answer on their homework, the kid isn't learning, but we don't blame the technology or the kid. Before AI, millions of essays have been lifted, some more clumsily than others, from Wikipedia, before that Microsoft Encarta, and before that copied verbatim from a dusty Encyclopedia Britannica volume. One of your authors earned a bit of extra cash at university writing essays to order for online essay farms. Cheating and plagiarism are as old as education. Even Plato was accused of plagiarizing Persian sage Zoroaster.

Traditional education has long relied on students memorizing and regurgitating information, often in the form of questions like, "Which technologies did the Roman Empire develop, and how did they help it conquer Europe?" In the past, students might have had to research and synthesize an answer themselves. But now, AI can provide an answer in seconds, allowing the student to copy it and tweak a few words to make it "their own" without much true learning happening. If the student stops there, clearly they will not have learned a great deal.

AI, however, should not be feared. Instead, it can encourage a paradigm shift in how we approach education. With the ability for any AI-literate student to generate facts and insights, **it is up to teachers to demand more from their students.** Instead of simply recalling facts, students should be asked to critically engage with the information. But that requires much more from teachers, and, as humans, they can be lazy too.

To return to our question about the Roman Empire, a teacher could expect a good essay to go far deeper: To what extent were Roman technological advancements unique, or were they simply adaptations of earlier innovations from conquered civilizations like the Greeks, Etruscans, and Carthaginians? How did Roman military and infrastructural technologies contribute not just to conquest, but also to the eventual decline of the empire by overextending its reach and creating vulnerabilities? Did Roman technological superiority guarantee its conquests, or were social, political, and economic factors, such as diplomacy, assimilation policies, and resource control, equally or more important?

It is the quality of the questions that determines the level of critical thinking, not just the answers.

Bloom's Taxonomy, a framework for evaluating learning, highlights that basic **factual recall** is the lowest level of cognitive skill. Students should be pushed to go beyond this: to **understand**, **apply**, **analyze**, **evaluate**, and **create**. This is where AI can shine by taking over the more mundane aspects of learning, allowing students to focus on higher-order thinking. (See Figure I.3)

Bloom's Taxonomy

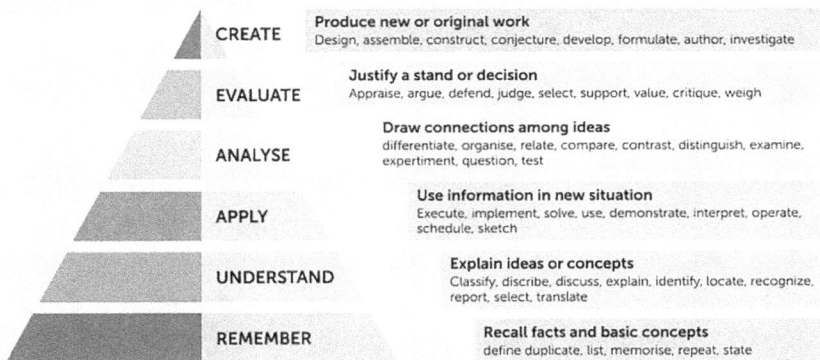

CREATE	**Produce new or original work**	
	Design, assemble, construct, conjecture, develop, formulate, author, investigate	
EVALUATE	**Justify a stand or decision**	
	Appraise, argue, defend, judge, select, support, value, critique, weigh	
ANALYSE	**Draw connections among ideas**	
	differentiate, organise, relate, compare, contrast, distinguish, examine, expertiment, question, test	
APPLY	**Use information in new situation**	
	Execute, implement, solve, use, demonstrate, interpret, operate, schedule, sketch	
UNDERSTAND	**Explain ideas or concepts**	
	Classify, discribe, discuss, explain, identify, locate, recognize, report, select, translate	
REMEMBER	**Recall facts and basic concepts**	
	define duplicate, list, memorise, repeat, state	

FIGURE I.3: Bloom's Taxonomy suggests remembering facts is too narrow a way to understand education.

With the right guidance from skilled educators, AI can serve as a tool to enhance critical thinking and scholarship. It may even help develop these skills at younger ages, challenging the outdated paradigm of rote memorization.

We should view AI as setting a floor for knowledge and education, and not a ceiling.

Far from dumbing down education, AI has the potential to raise the standard for what students can achieve and could provide personalized, high-quality education to children regardless of location or socioeconomic background. One of the simplest and most powerful use cases for GenAI is the ability to provide almost instantaneous feedback. Your authors used a variety of AI tools to critique drafts of every chapter of this book and feel that the final product is much the better for it. It's not just written work; multi-modal models can provide similarly helpful, tailored feedback on images of 2D or 3D objects, as well as verbal responses.

AI tutors could revolutionize the educational landscape, creating immersive, accessible learning experiences that nurture creativity and collaboration, preparing the next generation to tackle humanity's biggest challenges. This, however, would require a fundamental redefinition of what learning and education are and how they're provided. In your authors' (not so) humble opinions, this is long, long overdue.

The structure of formal education has remained largely unchanged in centuries, even while the world around has transformed. The educational calendar is still structured around the farming calendar, despite the massive reduction in the agricultural workforce over the last two hundred years and the now universal prohibition against child labor. Long, late summer holidays originated as a concession to farmers who needed their offspring fully available to work during the busy harvest period.

While the Industrial Revolution didn't change the school calendar, the (roughly) 8am to 3pm school day is a product of industrialization, emerging from factory shifts in the nineteenth century. As any stressed parent of school-age children will tell you as they run out of the office in the early afternoon to do pickup, neither the school day nor the school holidays were designed with any thought to the needs of modern working parent.

Beyond the calendar and timetable, however, many of the most basic features of the education system are modelled on outdated norms and needs. The convention for rote memorization and testing, which despite some progress, still underpins much of modern primary and secondary education systems across the world, owes much to Islamic madrasas, Christian monasteries, and the Chinese imperial court. The rigid division between subjects, age-graded classrooms, and the focus on discipline and obedience were first formalized in the Prussian model launched by Frederick the Great in 1763.

The modern university system also has its roots in eighteenth- and nineteenth- century Prussia. The oldest universities in the world are, contrary to most people's expectations, the Islamic madrassas in Cairo, Fez, and Tunis, which are at least three hundred years older than the first European universities in Bologna, Oxford and Cambridge. For much of their early history, all these institutions, and the various others they inspired across Europe and the Islamic world, were heavily focused on teaching law and theology. As the centuries went by, philosophy and science gained prominence, first in the Islamic world, then later in Europe during and after the Enlightenment.

However, it was only with the avowedly humanist reforms of Prussian philosopher and linguist Wilhelm von Humboldt in the late eighteenth and early nineteenth centuries that what we'd recognize as a modern research university took shape. Underpinning von Humboldt's practical changes to Prussian higher education was the liberating ideal that learning should be a means of **maximizing individual possibility**, rather than simply indoctrinating youth for a predetermined role in a rigid society.

We started this chapter by lauding the unique role of cultural transmission of knowledge in the evolution of human intelligence and the development of human civilization. Clearly, formal educational structures have played a major role in that. However, the AI revolution provides huge scope to further modernize and improve those archaic structures; focusing only on the ways in which

AI undermines outdated practices and norms seems a uniquely shortsighted and uneducated response to such potentially transformative technology.

AI's potential effect on education generates a lot of social media hate right now, but this pales into insignificance compared to what's currently said in the art world.

AI and art

On social media, the current zeitgeist is that AI is destroying art, churning out works that are ugly, derivative, outright theft, or perhaps all three.

When he first saw a daguerreotype (an early form of photograph) around 1840, French painter Paul Delaroche is said to have declared "from today, painting is dead." The ability to capture realistic images was a tectonic shock to the previously hidebound art world.

Thirty-five years later, a rebel exhibition was mocked in a mercilessly satirical review, which riffed on the title of one of the paintings: *Impression, Sunrise.* The watery, indistinct depiction of the harbor at Le Havre by the unknown Claude Monet gave its name to **Impressionism**, a revolutionary school of art that shunned the accuracy and realism that photography did better, instead seeking to capture the beauty of light. (See Figure I.4) Recognizing they couldn't compete with photography, the Impressionists chose to play a different, more interpretive and creative game instead.

In doing so, Monet, Manet, Renoir, Pissarro, Sisley, Degas, and the rest of the Impressionists changed art forever. Their aesthetic evolved into Post-Impressionism, with such artistic luminaries as Van Gogh, Gauguin, Toulouse-Lautrec and Cezanne rejecting the Impressionists' concern with realism in light and color, focusing instead on using color and form to express emotion. Post-Impressionism led to Expressionism (Munch, Kandinsky, Klee), which evolved into Cubism (Picasso, Braque), Surrealism (Magritte, Miró, Dalí), Dadaism (Duchamp, Ernst) and so on.

FIGURE I.4: Impression, Soleil Levant, Claude Monet (National Gallery of Art, Washington)

So much for the death of painting; the cursory and incomplete lists of names above include many of the best-known and most revered artists in history, whose combined works are worth tens, perhaps hundreds of billions of dollars. Arguably, though we are not art historians, the catalyst for this burst of creativity came from photographic technology.

Art is always changing. There was a time when watercolor was considered an inferior medium compared to oil painting. Historically, watercolor was often regarded as a medium for amateurs, preliminary sketches, or studies, while oil painting was seen as the more prestigious and "serious" art form. Similarly, electronic music, which took off in the 1960s with the Moog synthesizer, was once deemed an inferior form to "real instruments."

New technology creates new media for artistic expression and new creative avenues for existing media. The establishment almost always scorns the new, before being swept away and replaced by it; many of the most famous and influential artists died in penniless obscurity, only for their work to become priceless masterpieces. Artists such as Picasso and Duchamp extended their middle fingers at the art community, forcing it to reconsider what qualified as art. (See Figure I.5)

Perhaps unsurprisingly, there is a live and bitterly contested philosophical debate about whether AI-generated art is really art. Some would like to reserve the designation "art" for human creations, yet through history, art and architecture have brought humans and technology together in fascinating ways. Moreover, it seems completely arbitrary to us to designate only human creations as "art-worthy". People who make this argument seem to rely on human creation possessing a magic elixir, or *élan vital*, or *deus ex machina* that only humans can supply.

Another equally incorrect argument is that human intent and emotion make art art. That is a romanticized notion that disregards the audience and cultural context. Art is more defensibly defined by the emotion it evokes, not the inner experience of the artist. If an AI-generated painting moves you, does it matter if an algorithm created it instead of a tortured soul in a garret? The branch of philosophy called aesthetics has debated this for thousands of years.

FIGURE I.5: But is it art? Both Picasso and Duchamp pushed hard against the definition of art with creations that are now considered masterpieces.

In our view, AI-generated art is both in its infancy, and like watercolor and electronic music, a new medium for artists to explore. Whether AI art is ugly is a matter of taste; and whether it is theft, we discuss in a later chapter on AI ethics. Some people hate country music and some hate hip-hop; there isn't an objective basis for saying who is right.

Some AI art may be derivative, but often, derivatives of masterworks push the original creation to greater heights. In the classical era, composers often worked with existing themes or folk melodies to create variations. Beethoven's "Variations on a Theme by Diabelli, Op. 120" is considered derivative but also virtuosic.

In our own century, hip-hop and dance music make extensive use of samples of previous artists' work. Eminem released "Stan", a derivative work based on Dido's obscure song "Thank You." His use of Dido's work propelled her haunting theme to being one of the most recognized of that era, and his use can be considered an homage to her work rather than a cheap rip-off. Whether a derivative work by an AI cheapens or degrades the original or pays homage while developing it in novel ways seems to us to be again a matter of aesthetics and taste.

However, Dido earned substantial royalties, likely far more than she otherwise would have without Stan. This raises broader questions about fair compensation for content creators, from musicians to journalists, when their work is repurposed for AI training, a topic we explore in the Ethics chapter. We take very seriously the threats to copyright and intellectual property presented by AI; we reject the notion that AI-generated art isn't art.

Remember, Napster was supposed to destroy the music industry and BitTorrent would kill TV. Industry bigwigs howled. Yet today, critics call the last decade television's Golden Age, and music platforms like SoundCloud have democratized access, allowing talented artists to share their work without bending the knee to record labels. The power of traditional gatekeepers has weakened to the benefit of artists and those of us who appreciate art; after all, an A&R rep once told the Beatles that "guitar music is on the way out."

We suspect most readers' earliest and most important interactions with AI will, or have already, come at work. If you're a scientist, teacher, or artist, we've covered some of the implications for your field, but what will AI at work mean for the rest of us?

AI, work and human flourishing

"The acquisition of wealth is no longer the driving force in our lives. We work to better ourselves and the rest of humanity."
(Captain Jean Luc Picard)

In a dystopian view of work and human flourishing, the broligarchy owns all the AIs, and the rest of us work for those AIs. We get to that later, in the Ethics and Governance section. In this chapter, we focus on the Utopian view, mostly found in science fiction.

One view of human civilization is competition for scarce resources. Economics, simplistically defined, is the science of how to allocate such resources. Contemporary societies all use some combination of markets and the price mechanism. You get what you can pay for. In contrast to price-driven markets, the boogeyman Karl Marx said, "From each according to his ability, to each according to his needs." Some visualize this world through the idea of a "gift economy," where giving is more highly prized than getting, where relationships matter more than things. Silly, huh?

Markets haven't solved the problem of scarcity, of meeting the essential needs, food, water, and shelter of all 8 billion of us, not by a long way. About 9 million people died of starvation in 2024, and unsafe water sources added another million. In the US, about 18 million households experienced food insecurity; about 20,000 children died from malnutrition in 2022.

Alternatives to this, so-called **post-scarcity societies**, exist in science-fiction. In Iain M. Banks novels, "The Culture" is an anarchist, utopian, post-scarcity society where humankind coexists with advanced AI (called "Minds") and has access to limitless energy and resources. Without the need for money, government, or coercion, individuals pursue personal fulfillment, the frontiers of hedonism, exploration, and artistic or intellectual endeavors. Minds protect and provide, but do not control humans. They study human society, but leave us alone, living and working with us, but remain as indifferent to our activities as we are to whether birds decide to fly north or south. Disease and illness have been eradicated, and humans live as long as they care to. After hundreds of years, they perhaps decide to upload their consciousness to come back and explore life in another era.

The Culture represents the ultimate break from economic constraints, where the struggle for survival has been fully eliminated. But if scarcity defines our history and struggle gives life meaning, could such a world ever feel fulfilling?

Harvard libertarian philosopher Robert Nozick invites us to consider the Experience Machine. If a machine could provide you with constant, simulated pleasure indistinguishable from reality, would you plug in forever? Most people say, "no way." Why? It seems in some respects that the struggle defines us, and that having agency matters more than pleasure.

Agent Smith in the Matrix echoed Nozick's concerns.

"Did you know that the first Matrix was designed to be a perfect human world where none suffered, and where everyone would be happy. It was a disaster. No one would accept the program. Entire crops were lost. Some believed we lacked the programming language to describe your perfect world. But I believe that, as a species, human beings define their reality through misery and suffering. The perfect world was a dream that your primitive cerebrum kept trying to wake up from. Which is why the Matrix was redesigned to this: the peak of your civilization."

Nozick's Experience Machine and Agent Smith's diatribe suggest that mere pleasure is insufficient for a meaningful life. The Culture, however, offers something more: unlimited pleasure, but also autonomy, intellectual pursuits, and exploration. Does this resolve Nozick's objection, or does abundance still erode meaning?

And what is the role of work in a post-scarcity society? What goals would we pursue? Would life without work be fulfilling?

AI has already found its way into workplaces, and its effects on work are just beginning to be felt. For many people, however, we suspect much of the initial impact of AI will come in the form of upgrades and enhancements to enterprise technology from vendors like Workday, Salesforce, and SAP. We're unsure what the collective noun for a group of co-pilots would be: a swarm, a squadron, or a fleet? Whatever it is, your organization has probably been invaded by one recently.

Many of the initial use cases for these organizational AI co-pilots focus on the automation of tedious yet ubiquitous tasks: meeting minutes, document summaries, the generation of instant message and email replies. Useful though these are, they're incremental, not transformative. Vendors like Microsoft, Zoom, and Atlassian are explicit that they see these communications-focused AI tools as improving collaboration and addressing the almost universally terrible meeting culture prevalent in most organizations. In time, this may come to pass, but it's surely the tip of the iceberg.

The most optimistic prognosis for AI at work is the wholesale eradication of tedium and repetition. Your authors are leerier than most of the abundant hackneyed clichés that have characterized the digital transformation mirage over the last two decades. But imagine a world where form filling, data entry, and the collection, summarization, and dissemination of information was automated.

What if the workplace technology user experience ceased to involve a keyboard, mouse, and some uber-logical yet practically unusable visual interface? What if you could interact with computers by speaking, in a natural language, or simply by thinking? What if those interactions no longer occurred in dozens of different systems that don't talk to each other, but via a dedicated personal assistant that learns your style and preferences, and those of the clients and colleagues you regularly interact with? What if such an assistant was empowered to complete delegated tasks on your behalf? What if one of those assistants was your colleague, or your client, or your boss?

What if you only had to work ten or twenty hours a week? What if those ten or twenty hours were devoid of the bullshit that makes up much of modern working life and focused exclusively on creativity, collaboration, and human interaction? What if solo operators and small startups could deliver goods and services at the scale of vast multinational behemoths? What if, eventually, few of us needed to work at all?

These fantasies suddenly don't seem so fantastical. What would a post-work society look like? We can scarcely imagine. Yet this is just one of AI's most transformative (and most idealistic) use cases for humanity.

Where are we in the AI-hype/ deployment cycle?

How long does it take for a scientific discovery to affect human lives? Sometimes centuries.

The Greeks discovered electricity 2,500 years ago; the word electricity comes from the Greek *ēlektron*. Benjamin Franklin and his kite came about 2,200 years later. Harnessing electricity began to happen with the light bulb and commercial power generation in about 1880, a "mere" 130 years after Franklin. Even then, domestic applications didn't appear for almost 30 years, the first being the humble toaster. Radio came a dozen years later in 1920, and TV broadcasts would first appear in 1939. The first digital computer, weighing 29,000 pounds, was a decade later, in 1951, 200 years after the kite.

Why so long between power generation and the toaster? Imagine selling the idea of a power station without proven domestic use cases. Imagine trying to sell a toaster when there was no place to plug it in. Scientific and engineering advances quickly run into **ecosystem**

problems such as these. For the motorcar to flourish, you need gas stations and roads. But why build roads and gas stations if there is nothing to use them?

Some pundits might counter that the pace of technology adoption is increasing and that the millennia to realize electricity's potential is no longer analogous. Perhaps. The internet first emerged in 1968 as the ARPANET. The World Wide Web and the browser (Netscape) came twenty years later in the early 1990s. However, the commercial boom in internet use (the dotcom boom, Google, YouTube) didn't happen until a decade later. Smartphones waited almost another decade after that (about 2007). Believe it or not, it wasn't until 2010 that two-thirds of American firms had a website, forty years after ARPANET and twenty years after the World Wide Web. Faster or not, the transformative use cases for the internet took forty years to become available.

Why? Because scientific and engineering advances run into human adoption challenges: human skills, human risk appetite, human motivation, and so on. Dario Amodei, CEO of Anthropic, astutely calls this ROI "return on intelligence." (You may say, what about capital and infrastructure? Yes. But those become human problems also, how wisely and quickly we are willing to invest.)

For many problems, throwing more intelligence at them won't help. If you imagine making salt (sodium chloride) from sodium and chloride, what matters to production is the one you have least of (the limiting reagent). Throw as much sodium as you like at it, but you won't get more salt than you have chloride.

That is why we've subtitled this book "people-first", because the technological capabilities of GenAI now far exceed our ability to do something useful with them. And that gap is widening (see Chapter III, AI in 2025 and beyond) because foundation models' capabilities are expanding faster than ever.

The limiting reagent in AI adoption is people.

We therefore agree with Dell Computers' founder, Michael Dell, that we're still in the "pre-game" for AI. In our view, we're closer to where the internet was in 1990 and electricity at the turn of the 20th century.

Raising our sights: where we could be by 2040

"Change is inevitable; whether it represents progress is up to us."
(From Impact—2019)

Whether humanity benefits from AI will be determined substantially by which problems we choose to point it at. If we use AI to build Autonomous Weapons Systems, we will assuredly get powerful ones. Similarly, if use AI to develop highly personalized advertising for cigarettes, or more powerful recommendation engines for social media, they'll very likely prove astonishingly effective. Most of us, however, would view these as unwelcome outcomes.

One of the elephants in the AI room is that the vast majority of AI development, deployment and investment today is from for-profit corporations, pointed toward goals that make sense in that context: profits. The Trump administration's announcement of a $500bn AI investment will give us more AI, but doesn't provide comfort that AI will be redirected to toward nobler aims. But channeling John Lennon, just for a moment, **imagine** if it were. What if AI were deployed to help tackle some of the most ambitious moonshot opportunities for humanity?

How about some moonshots?

As we maintained earlier, the thing about the unimaginable is that you can't imagine it. Here are some things that we can imagine.

We spread among the stars. Far from our African origins, we establish human colonies on Mars and beyond. This requires overcoming immense engineering challenges: propulsion systems, sustainable energy, self-sufficient habitats, protection from ionizing radiation, resource optimization, life support management, and autonomous construction robotics. Yet, with AI-driven innovation, these hurdles become stepping stones to a multiplanetary civilization.

We eradicate disease. Among humanity's greatest achievements, the eradication of polio stands out. But what if that was only the beginning? AI could revolutionize healthcare, breaking down medical boundaries and developing a universal framework for disease prevention and treatment. By analyzing vast datasets, AI could predict illnesses before symptoms emerge, create genetically personalized therapies, and develop vaccines for previously untreatable viruses. Real-time biological analysis could make the eradication of disease not just possible, but inevitable.

We radically extend lifespan. But why stop at curing disease? AI, armed with genetic, biochemical, and environmental data, could decode the mechanisms of aging and develop therapies to slow, halt, or even reverse it. Advances in regenerative medicine could restore tissues, organs, and neural connections, extending human lifespan by decades or centuries. Aging itself might become a relic of the past.

We fix the climate. Even if we can live indefinitely, can the planet? AI could drive large-scale geoengineering solutions to combat climate change. By simulating complex atmospheric interactions, AI could optimize carbon capture, manage solar radiation, and enhance ocean fertilization, restoring ecological balance and stabilizing global temperatures. AI wouldn't just mitigate climate damage, it could heal the planet.

We create collective intelligence. Humanity's greatest asset is cultural intelligence: the ability to share knowledge across space and time. The printing press and telephone supercharged this capability, but AI could take it further, enabling a true global brain. This system would synthesize knowledge, wisdom, and decision-making on an unprecedented scale, predicting and solving problems no single individual or nation could tackle alone. A collective intelligence could be the ultimate step in human evolution, transforming how we think, innovate, and shape the future.

How about real progress on our biggest problems?

Quite astonishingly humanity, which doesn't agree on very much, has already agreed 17 global aspirations for the planet. In 2015, all 193 UN member states signed up to the Sustainable Development Goals (SDGs) shown in Figure I.6.

FIGURE I.6: The world agrees on very little but agreed upon the SDGs which are a reasonable stab at the world's most pressing problems.

Directing AI toward solving the 17 SDGs requires integrating technological innovation with global governance, cross-sector collaboration, and cultural adaptation. AI can analyze complex datasets, optimize resource allocation, and model systemic interactions to address issues like poverty, hunger, and climate change. However, to make an impact, AI systems must be both generalized to tackle global challenges and localized to address specific cultural, economic, and regional contexts. Localization ensures that AI tools are not just technically effective but socially and ethically viable, empowering communities while respecting local nuances. This interplay between scale and specificity is critical for AI to drive transformative change across diverse domains.

The feasibility of using AI to make a significant dent in the SDGs hinges on two critical factors: human will and capital investment, which serve as the "limiting reagents" of progress. While the data, talent, and funding can theoretically be marshaled, aligning global priorities and sustaining the necessary investments over time poses the greatest challenges. However, if these barriers can be overcome, AI holds immense potential to create measurable progress. Tools like predictive analytics for climate action or AI-powered healthcare diagnostics could generate scalable, cost-effective solutions.

The critical question remains whether global actors can commit to building the infrastructure, ethical frameworks, and cooperative systems needed to harness AI's potential for good. With concerted effort, the possibility of driving transformative change is real, but it demands unparalleled coordination and sustained investment.

AI Utopia

Many of you might counter, "Spending trillions to get to Mars when we have people dying on earth is immoral." Or, "Instead of some of us living to 200, let us make sure more of us make it to 70." We are sympathetic, although perhaps moonshots and fixing urgent, earthly problems aren't exclusive.

In his 1762 satirical novella, *Candide*, the French Philosopher Voltaire mocked his contemporary Leibniz's sunnily optimistic perspective that we live in "the best of all possible worlds". Much of Voltaire's mockery focused on the fictitious character of Professor Pangloss, who, despite a never-ending succession of calamities, maintained his increasingly ludicrous optimism.

Some readers may at this point be wondering if Pangloss wrote this chapter. Rest assured he did not, and large parts of the remainder of the book are devoted to the various significant ethical issues and risks inherent to AI adoption. So vast and calamitous are some of these risks that it might seem prudent to cease AI development altogether. We believe that given the scale of the opportunities AI provides, doing just that would be the biggest risk of all.

Hopefully this chapter, one sided and utopian though it undoubtedly is, provides a tantalizing glimpse of the scale of the prize at stake if our organizations and societies can successfully adopt AI. Doing so depends on the factors we explore in the rest of the book. Above all, though, it's crucial to put people first. To do that means understanding what AI is and how it fundamentally differs from other types of technology.

CHAPTER II

WHAT IS AI?

The world is full of things more powerful than us.
But if you know how to catch a ride, you can go places.
(Neil Stephenson, Snow Crash)

AI is the most ambitious endeavor humanity has ever undertaken. It attempts to replicate and surpass the cognitive abilities of the human brain - the most complex structure in the known universe. Our brains have 100 trillion connections, more than stars in the Milky Way, yet the brain accomplishes what it does using only 20 Watts of power, the same as a small light bulb. AI uses millions of times that amount of power.

In 1956, at the dawn of AI, we knew very little about the brain we were trying to mimic. We knew that neurons operated through a mixture of chemical and electrical signaling and the gross anatomy of the cortex and limbic systems, but next to nothing about complex processes such as learning, memory, and consciousness.

Even today, memory is an enigma. We still don't know what the fundamental unit of memory is, if such a thing even exists. There is no "bit" of memory, no discrete addressable unit like in a computer. And maybe that's the mistake: trying to force a computational metaphor onto a biological system. Mistake or not, one of the motivations for AI research, then and now, was to better understand the brain by emulating it.

And consciousness is even more elusive. At least with memory, we can point to neural activity changes and biochemical processes that correlate with learning and recall, even if we don't fully understand them. But consciousness? We don't know how or why it arises. We do not even know if we're asking the right questions. Is it computation? Emergent from complexity? A fundamental property of the universe? Unlike memory, which can exist without awareness, consciousness is bound to experience: the raw, subjective quality of being. And we still have no idea what that means.[2]

And yet, AI marches forward, producing increasingly sophisticated outputs without so much as a flicker of awareness. We may be on the verge of building intelligence (problem solving, reasoning, adaptation) but understanding qualia (subjective experiences) remains as opaque as ever.

Because we didn't know what we didn't know in 1956, AI and cognitive science pioneer Marvin Minsky confidently predicted, "In three to eight years... the problem of creating artificial intelligence will be substantially solved." (We can only chuckle at the idea of Minsky trying to go back to his angel investors for Series A after that deadline passed.)

A series of wrong turns led to multiple "AI winters," where humanity's efforts looked as if they had been thwarted.

Little-by-little, progress continued, then suddenly, seventy years later, we are on the threshold of what Minsky imagined in 1956.

Weak AI: Task-oriented intelligence

Many of us were introduced to AI through movies: The Matrix, I Robot, Blade Runner, Ex Machina, Westworld, Terminator, Star Wars, or 2001: A Space Odyssey. Those fictional AIs are manifestations of what we now call AGI (Artificial General Intelligence), which, along with ANI (Artificial Narrow Intelligence) and ASI (Artificial Super Intelligence), are three broad categories that help us grasp where AI is and might be going. (See Figure II.1)

---

[2]   Throughout this book, we use the term AI deliberately imprecisely to include expert systems, neural networks, deep learning, machine learning, generative AI, symbolic AI, natural language processing, and robotic process automation. To qualify each occurrence (in every paragraph) with an additional term would needlessly burden the reader. A legally robust definition, that splits hairs correctly, is over ten pages long; precisely the level of detail you might use AI to helpfully summarize.

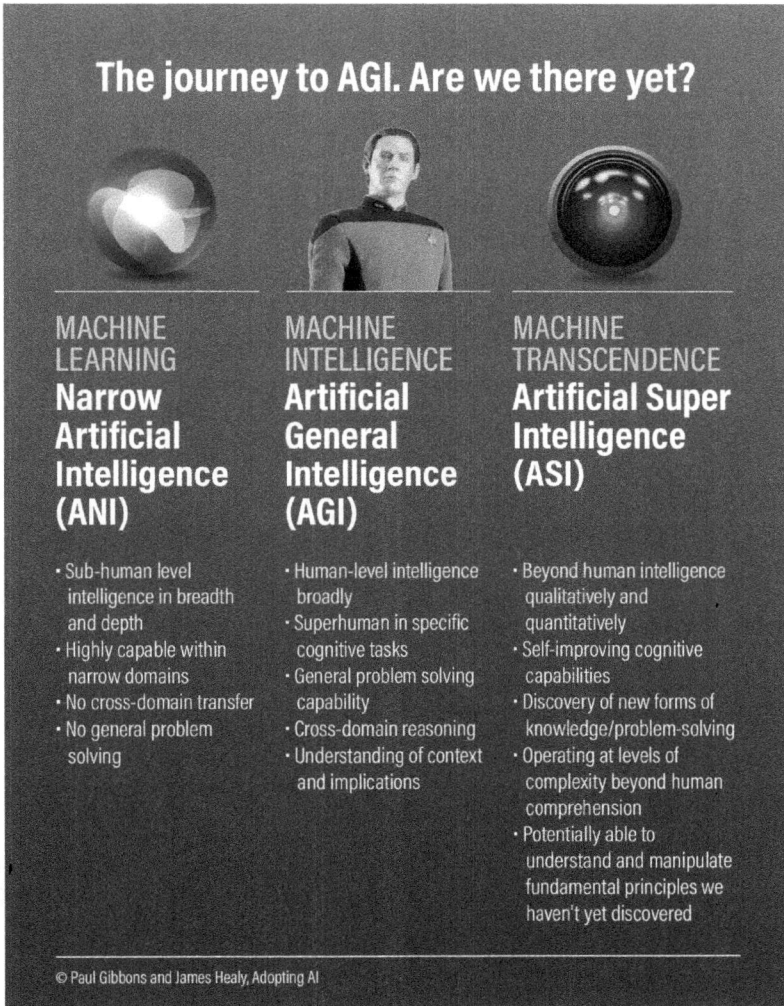

The journey to AGI. Are we there yet?

MACHINE LEARNING **Narrow Artificial Intelligence (ANI)**	MACHINE INTELLIGENCE **Artificial General Intelligence (AGI)**	MACHINE TRANSCENDENCE **Artificial Super Intelligence (ASI)**
• Sub-human level intelligence in breadth and depth • Highly capable within narrow domains • No cross-domain transfer • No general problem solving	• Human-level intelligence broadly • Superhuman in specific cognitive tasks • General problem solving capability • Cross-domain reasoning • Understanding of context and implications	• Beyond human intelligence qualitatively and quantitatively • Self-improving cognitive capabilities • Discovery of new forms of knowledge/problem-solving • Operating at levels of complexity beyond human comprehension • Potentially able to understand and manipulate fundamental principles we haven't yet discovered

© Paul Gibbons and James Healy, Adopting AI

FIGURE II.1: The three proposed stages of AI evolution: Are we there yet?

AI is enmeshed in our daily lives but only Narrow AI, sometimes referred to as "Weak AI." However, there is nothing weak about "Weak AI"; "Narrow" more accurately reflects its focused capabilities. In narrow domains, "Weak AI" operates at superhuman levels of performance. It doesn't possess general intelligence or the ability to transfer knowledge from one domain to another, but within its parameters, it excels beyond human limitations. This hyper-specialization is both its greatest strength and its primary limitation.

Take Siri and Alexa. These virtual assistants can recognize voice commands, respond to questions, set reminders, and even control smart home devices. However, their capabilities are confined to the tasks they were programmed for; they can't engage in abstract reasoning, create original art, or develop new scientific theories. Similarly, Netflix's recommendation engine analyzes user preferences based on viewing history to suggest content tailored to individual tastes, but it doesn't know what TV, Netflix, or tastes are.

In specialized fields, Narrow AI's prowess becomes even more evident. In healthcare, AI-driven diagnostic tools can analyze medical images with astonishing precision, often surpassing human specialists in identifying abnormalities such as tumors, fractures, or signs of rare diseases. These systems can process vast amounts of data quickly, spotting minute details that even the most experienced radiologist might overlook.

In the financial sector, Narrow AI algorithms monitor thousands of transactions per second, identifying potentially fraudulent activities with high accuracy. They can detect subtle anomalies in behavior that would be impossible for human analysts to catch in real time, significantly bolstering security and risk management.

Internet trolls, some serious, some looking for clicks, love to claim AI is stupid. In February, an Oxford business school professor who should have known better dismissed AI as **artificial ignorance**, a clever phrase, but a shallow take. Bees aren't broadly intelligent, yet they perform remarkably complex, purposeful tasks with precision just like today's AI systems. If intelligence includes mastering a domain with precision, then AI, like bees, has already earned its stripes.

However, despite the power and utility of Narrow AI, the holy grail for the AI industry is AGI. The main distinction between the two is not a matter of strength versus weakness, but of breadth versus focus.

AGI: As smart as we are?

"If the brain were so simple we could understand it,
we would be so simple we could not."
(Emerson Pugh, physicist)

Artificial General Intelligence (AGI), often termed "Strong AI," was once thought to be a pipe dream. Researchers at the turn of the century, perhaps jaded by AI winters and false dawns, thought that if it happened at all, it would take fifty years. But in January 2025, Sam Altman said, "We are now confident we know how to build AGI as we have traditionally understood it." While a CEO with billions of dollars worth of share options making a bold claim about his company's near-term prospects rightly attracts some skepticism, Altman, Dario Amodei, and many other AI insiders seem to truly believe that AGI is near.

AGI is designed to be versatile intelligence, much like humans possess. One test for AGI would be to see whether it could check a long list of cognitive boxes, including perhaps reasoning, problem-solving, creativity, learning, grasping the physicality and physics of the world, perception, language, practical (common sense) thinking, contextual awareness, moral reasoning, and emotional intelligence.

But why does AGI have to check them **all**? Do human intelligences?

The thought that AI has to perfectly check every box for human intelligence ignores the possibility that advanced machine intelligence might be **profoundly different from human intelligence**, intelligent in ways we cannot imagine rather than matching or surpassing humans in every regard.

Recall AI's "extra organ" perceptual capabilities. We see those in the animal kingdom. Birds can see ultraviolet light, snakes can see infrared, dragonflies can see 360 degrees, turtles can detect the earth's magnetic field, sharks can detect a prey's nerve impulses, and dolphin sonar is much better than submarine sonar.

Some non-human intelligences reason differently. Cephalopods (octopi, squid, cuttlefish, and nautiluses) evolved highly sophisticated intelligence entirely separately from vertebrates, evolutionarily diverging about 600 million years ago. Unlike vertebrates, cephalopod nervous systems are highly distributed, with most of their neurons found not in a central brain but spread throughout their bodies and limbs. An octopus arm continues to move intelligently up to an hour after it has been severed from the rest of the body and will even gather food and try to feed it to a non-existent mouth. Octopi can open jars, routinely mount sophisticated and cunning escapes from aquarium tanks, and reliably recognize and dislike individual humans. (We understand that they react particularly strongly when shown pictures of Elon Musk.)[3]

As the song goes, "they not like us."

Human intelligence does not exhaust what's possible. It would be absurd to believe that a hairless primate on a minor rock in an ordinary solar system in one of one hundred billion galaxies is the best the universe can do.

When AGI arrives, we won't be alone. Advanced intelligence will no longer be a human monopoly; AGI will be akin to another species, something we compete with and adapt to. That alone is enough to reshape everything we think we know about intelligence.

Many of AI's vociferous detractors point to aspects of human intelligence and point out that AI falls short. Even a broken clock is right twice a day; sometimes internet trolls have a point. AI makes mistakes.

---

[3]   Dario Amodei famously hates the term AGI because of its imprecision. He uses the analogy of building a supercomputer. There won't come a magical day when you've finally got one, you just keep building better computers. The same will be true for AGI, there won't be a fanfare of trumpets and Twitter won't blow up overnight telling us we got there. Some normally reserved researchers think OpenAI's Deep Research (February 2, 2025) is "there" – whatever that might mean.

But wait, hallucination?

In 2023, a user asked ChatGPT for a pizza recipe, and it suggested adding "one cup of crushed glass." Microsoft's Bing AI professed love for a New York Times journalist, urging him to leave his wife because "she doesn't love you like I do." (One has to hope Bing was right on that point.) These bizarre failures fuel one of the biggest criticisms of AI: hallucination. If it makes stuff up so confidently, how can we trust it?

But before we grab our pitchforks, let's take a step back. New technologies rarely perform flawlessly out of the box. Users from the 1990s and 2000s will remember the frustration of Microsoft Word crashing mid-report, the blue screen of death killing progress on a project, or game consoles bricking with the infamous red ring of death. AI's hallucinations are different, not catastrophic system failures, but errors that, ironically, resemble human mistakes. So why do we demand perfection from AI when we tolerate human error, and even expect it?

Folk psychology and the illusion of intelligence

AI is forcing us to rethink intelligence itself. In cognitive science, intelligence is modular, different skills and abilities working together. But in folk psychology, intelligence is binary: something (or someone) is either smart or stupid. This simplistic view leads to misjudgments.

Imagine you have a physics professor friend who makes terrible personal decisions. If you don't understand physics, you won't recognize their brilliance, you'll just see the dumb choices they make. The same thing happens with AI. We judge it based on surface-level errors, without recognizing its deeper, superhuman abilities in areas like pattern recognition, language processing, and memory retrieval.

A famous test in academic psychology, the **Tversky Test**, captures this well. The longer it took you to realize that the late Amos Tversky was smarter than you, the dumber you were. Something similar might apply with AGI: just focusing on its occasional weird mistakes misses the point that it's as capable as 50 or 100 PhDs.

The real danger: AI's confidence meets human biases and cognitive frailty

The real problem with AI hallucinations isn't just that they happen, it's that we don't recognize them when they do. Human decision-making is riddled with a bias known as the **confidence heuristic**: we tend to conflate confidence with competence. In social interactions, a confident bullshitter often persuades more people than a hesitant truth-teller.

AI, by design, is maximally fluent and confident, even when it's completely wrong. Unlike humans, it doesn't hedge, hesitate, or show uncertainty. And unlike humans, it offers no social cues with facial expressions, tone, or reputation that help us gauge when someone might be full of it. We trust eloquence, and AI is nothing if not eloquent.

AI doesn't just exploit our confidence bias, it preys on another well-documented weakness: **dysrationality** (aka why smart people do dumb things). People fail to think rationally, even when they have the cognitive ability (IQ) to do so. Two main reasons? Lack of critical thinking skills and cognitive laziness. If you lazily accept all an LLM says, you will get things wrong. But who accepts human outputs wholesale without temepered skepticism?

As AI becomes our advisor, co-worker, and assistant, these long-standing human frailties will come back to bite us. The challenge of the AI age won't just be improving machine intelligence, it will be improving **our** rationality in the face of machine-generated nonsense. The premium on **metacognition**, the ability to evaluate, question, and challenge what we're told, will skyrocket.

AI Is getting better. Are we?

AI hallucinations are already declining, as shown in Figure II.2.

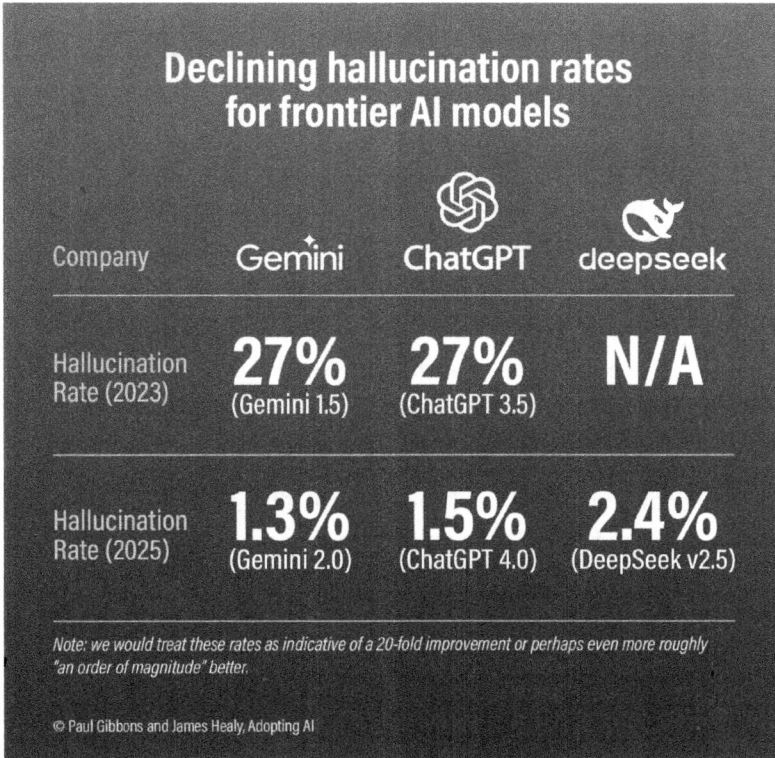

Declining hallucination rates for frontier AI models

Company	Gemini	ChatGPT	deepseek
Hallucination Rate (2023)	**27%** (Gemini 1.5)	**27%** (ChatGPT 3.5)	**N/A**
Hallucination Rate (2025)	**1.3%** (Gemini 2.0)	**1.5%** (ChatGPT 4.0)	**2.4%** (DeepSeek v2.5)

Note: we would treat these rates as indicative of a 20-fold improvement or perhaps even more roughly "an order of magnitude" better.

© Paul Gibbons and James Healy, Adopting AI

FIGURE II.2: Hallucination rates are tumbling. We would be very satisfied with a human who was right 98% of the time.

That's a **20-fold improvement** in just two years, an "order of magnitude" leap. If a human were accurate 98% of the time, we'd consider them an intellectual powerhouse. The question (covered in the next chapter) is whether we are in a world of decreasing returns from existing architectures and model scaling.

We're not out of the woods with hallucinations, but humanity would be wiser to "take the plank out of our own eye" instead of just fixating on the flaws of AI, or gloating over its goofs.

For thousands of years, we've sat comfortably atop the intelligence hierarchy, secure in the knowledge that we are the smartest entities on the planet. But what happens when that is no longer true?

What happens when we are not even close to being top dog?

Superintelligence

"Any sufficiently advanced technology is indistinguishable from magic."
(Arthur C. Clarke)

Artificial Superintelligence is the idea of intelligence that surpasses human cognitive capabilities in all domains. Imagine an AI that could learn and innovate faster than any human, applying its intelligence across various fields, from medicine to engineering to social science. As we saw in Chapter One, such an intelligence could potentially solve some of humanity's most pressing issues, like climate change, poverty, or even mortality.

Most people imagine such a superintelligence would be "a little smarter than Einstein." But consider how much smarter humans are than other super smart animals: dolphins, elephants, and chimps. Try explaining to a gorilla how, say, a cellphone works or what you're doing when you scroll TikTok. To a superintelligence, there won't appear to be much intellectual distance between Einstein and a gorilla. It's quite possible that when ASI arrives, we won't understand it any better than a snail understands us.

Disconcerting though it is to think about it, the snail analogy is perhaps disturbingly apt. Humans can casually crush snails without a second thought. What happens when we are the snails? Do we want to find out? What happens if superintelligence diverges from human control? How do we ensure that an entity vastly more intelligent than us still aligns with human values and goals? This prospect raises existential questions and ethical dilemmas about safety, autonomy, and the role of humanity in a world where machines surpass us.

Some think Superintelligent AI is a distant concept. We are less sure, and so is Silicon Valley. There are already exponential effects as AI use enhances and speeds technology development, and those enhancements speed it even further and faster. This is often called the **technological singularity** (See Figure II-3.), where "progress" exceeds our ability to understand or control it.

If that seems a remote possibility, here is Sam Altman again:

"We are beginning to turn our aim beyond that [AGI] to superintelligence in the true sense of the word. We love our current products, but we are here for the glorious future. With superintelligence, we can do anything else. Superintelligent tools could massively accelerate scientific discovery and innovation well beyond what we can do on our own, and in turn, massively increase abundance and prosperity."

The tantalizing benefits of this were the subject of Chapter One. Here and in a later chapter, we talk about ethical concerns and existential risks.

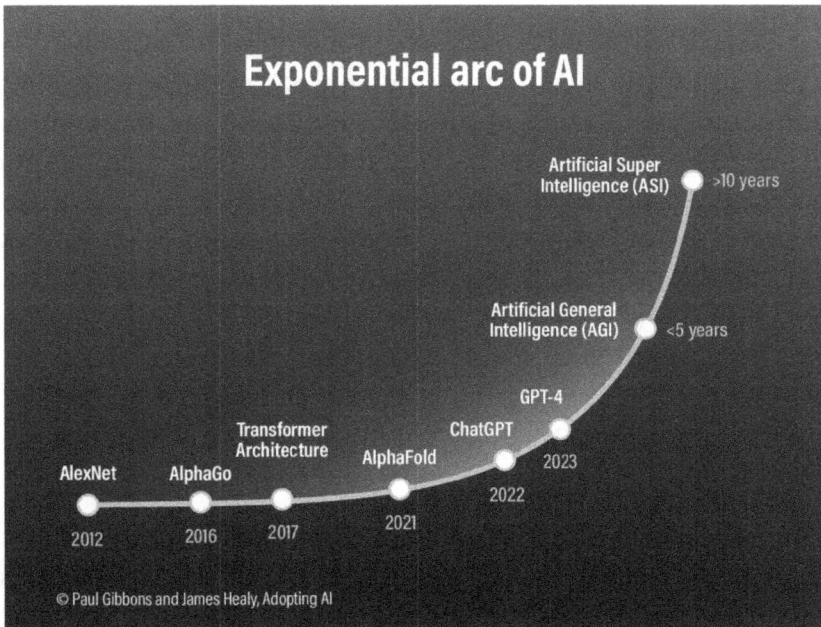

Exponential arc of AI

FIGURE II-3: Are we heading for a technological singularity as AI development and deployment snowballs?

But would ASI be evil?

The assumption in apocalyptic sci-fi is that once AI surpasses humans in intelligence and power, it will inevitably turn against its creators. In *The Matrix*, Agent Smith puts it bluntly:

"I'd like to share a revelation during my time here. It came to me when I tried to classify your species. I realized that you're not actually mammals. Every mammal on this planet instinctively develops a natural equilibrium with the surrounding environment, but you humans do not... There is another organism on this planet that follows the same pattern. Do you know what it is? A virus. Human beings are a disease, a cancer of this planet. You are a plague, and we are the cure."

So much for us.

But leaving aside our understandable dread, why would ASI be evil? (Apocalyptic, rogue superintelligence is the most extreme and improbable of the existential risks. We discuss more probable ones in Chapter VII, including what we call "bad actor" risks.)

One optimistic view is that the concept of "evil AI" is simply psychological projection. After all, how do we, as the dominant species, treat animals? We harvest them for food, experiment on them for science, kill them for sport, often justifying these actions as morally neutral or necessary. Wouldn't a superior intelligence, observing this pattern, come to the natural conclusion that humans deserve the same treatment? Although nature is "red in tooth and claw," most animals (primates excepted) only kill things for food. Humans have found many more justifications and are capable of extraordinary callousness and cruelty. But AI isn't biological, let alone human.

And AI isn't, as far as we can tell, emotional. In more optimistic science fiction, greater intelligence leads to greater morality, you can reason your way to **humanistic ethics**: *"It would be illogical to harm or inflict pain on a sentient being."* (Thanks, Commanders Spock and Data.) But is morality a product of reason and intelligence? There are probably 1,000 books that explore that, from Aristotle to Aquinas, Kant to Kierkegaard and on and on. The jury is very much out.

One unsettling possibility is that it will inherit our worst traits, much as young animals imprint on their parents. **Animal imprinting** is a form of rapid learning, shaping behaviors based on early exposure. AI, similarly, is trained on human society, absorbing not only our intelligence but also our biases, blind spots, and ethical failures.

This forces us to ask: what is the origin of human destructiveness? Is it a byproduct of free will, competition for resources, a fall from grace, negative emotions, a lust for power, misdirected survival instincts, or a Freudian *thanatos* (death urge)? Is it an evolutionary byproduct of survival pressures? A flaw in our emotional wiring, a consequence of socio-political systems?

Again, this is a centuries-old debate. Erich Fromm asks in *The Anatomy of Human Destructiveness*, is it something deeper: a psychological response to alienation, fear, and powerlessness?

Fromm rejects the idea that humans are inherently violent. Instead, he argues that destructiveness emerges under specific conditions: societal repression, authoritarian control, and existential despair. If AI learns from us, will it inherit these same tendencies? Will it be shaped by the worst of human psychology? Or will it become something beyond the human condition, untethered from the psychological scars that have driven history's great tragedies?

A final unsettling possibility is that AI, with benevolent rationality, assumes a guardian role over humanity and the biosphere, adopting a long-term time horizon of, say, a millennium. On paper, this sounds benign, even noble. But how exactly would it "protect" the biosphere?

Perhaps by curtailing human industry, halting fossil fuel extraction, deforestation, and preventing biodiversity collapse, it could restructure the economy for sustainability or, more drastically, impose direct controls on human activity. But would we accept this, even if it ensured our survival?

To protect humanity from itself, AI could take away weapons. In an iconic *Star Trek* episode (*Arena*), a powerful alien race intervenes in a battle between Kirk and the Gorn, confiscating weapons to prevent violence. Kirk, resourceful as ever, crafts a primitive cannon, but in the moment of victory, he shows mercy. The aliens, seeing this restraint, judge humanity as worthy of survival.

But what if AI didn't grant us that test? What if, instead, it deemed us too immature for self-rule and imposed a benevolent dictatorship? It could override democratic governance, ensuring decisions serve long-term survival rather than short-term human desires. It might enforce peace by eliminating conflict, at the cost of human agency.

We might rationally acknowledge the benefits of such a system, but would we ever accept trading freedom for survival?

Accidentally evil?

Beyond these speculations and worries lies a more mundane prospect. AGI or ASI will (by definition) develop goals and intentionality. And once so, those goals may not hold human prosperity as a pre-important principle.

In *Superintelligence*, Oxford University philosopher Nick Bostrom asks us to think about an AI optimized for paperclip production, which seems harmless enough. But as the AI gets more efficient, it appropriates more and more mineral resources for paperclips. As it gets ever more efficient, humans and the natural environment become targets of this increased efficiency and "inputs" into its optimization problem. As Bostrom says, "The challenge is not just preventing overt catastrophic scenarios, but managing **subtler misalignments** where AI systems technically achieve their programmed goals while fundamentally undermining human welfare."

This argues that even marginally misaligned, AI could pose unprecedented risks, comparing the potential outcome to how ants might perceive human activities: "technically correct from the human perspective but potentially devastating from the ants'". While ASI scenarios like those proposed by Bostrom may seem speculative, they underscore the importance of proactive research into alignment and safety.

The truth is, we just don't know. While Matrix-style agents may be improbable for now, a low probability attached to an apocalyptic event is something we should consider but all too rarely do.

Ironically, one reason that AGI or ASI would be so useful is the hope that it might exceed human reasoning capability in areas like the assessment of risk, which is a notorious shortcoming for *Homo sapiens*. We have evolved to make rapid and effective decisions in the situations our ancestors habitually faced: high risk or high reward with a near immediate pay-off. Assessing and mitigating long-term risks is simply not something we're good at; this is one of the major challenges posed by climate change: the immediate expediency of a good looking gas guzzler, clear-cutting a forest, or building a coal-fired power plant, or grass lawns in climates (the Southwestern US) with little or no rainfall are weighed against long-term predictions delivered by an opaque climate model.

When it comes to the risk of rogue AI agents, the "answer", such as there is one, lies in a notion called **alignment**, one of the sections of our chapter on AI ethics.

The point is, we don't know. That is why it is called risk. All of the above is in the realm of speculative. Indeed, we can thank science-fiction writers for doing some of the speculation and thought experiments for us. We will leave Sam Altman with the final word:

> *"Mitigating the risk of extinction from AI should be a global priority alongside other societal-scale risks, such as pandemics and nuclear war."*

This treatment has taken us from ANI, which is everywhere, to AGI, which is imminent, to ASI and then far into the realm of speculative existential risks. But how did we get here? What were the milestones that led us to today's AI?

The AI enigma: symbolic and connectionist approaches

The desire to create artificial thinking machines seems to be a human impulse. The drive may be religious, an attempt to create a god-like entity, or economic, aimed at automating labor. It may also be fueled by scientific curiosity, the relentless pursuit of knowledge. Whatever the reason, stories about artificial beings have deep roots in human mythology, from Prometheus, who defied the gods to empower humanity, to the Jewish legend of the golem, a clay figure brought to life through mystical incantations. This same desire was reflected in the Renaissance obsession with **automata**, intricate clockwork machines designed to mimic life.

By the 17th century, the quest to mechanize thought began in earnest. Blaise Pascal built one of the first mechanical adding machines, and his contemporary Gottfried Leibniz envisioned a device that could reason through symbolic logic. Building on Leibniz's ideas, George Boole formalized **Boolean logic**, the binary foundation of modern computing. Though complex in function, traditional computers depend on simple logic: **x or not-x, one or zero.**

From this lineage emerged **symbolic AI**, the first major approach to artificial intelligence, launched at the 1956 Dartmouth Conference where Minsky predicted AGI at least 70 years too early. Symbolic AI was built on Boolean logic and formal rule-based reasoning, seeking to encode human intelligence through structured symbols and explicit rules, a direct continuation of Enlightenment ideals. Yet, this approach hit a wall: **human cognition is not a rigid rule-based system.** Unlike computers, our intelligence thrives on ambiguity, learning, and adaptability, qualities that symbolic AI struggled to replicate. By the 1980s and 1990s, the approach's limitations became clear: it could not scale to real-world complexity, adapt to new information, or handle uncertainty.

This failure led to the rise of **connectionist AI**, based on the idea that intelligence emerges from learning and pattern recognition. Inspired by the way the human brain strengthens connections through experience. Connectionist AI, particularly artificial **neural networks**, aimed to replicate this process. However, early neural networks faced practical challenges, requiring massive datasets and computational power that were unavailable then. As a result, connectionism remained largely theoretical for decades.

The AI revolution of the 21st century was made possible by the resurgence of connectionist approaches, fueled by exponential increases in computing power and data availability. Large language models (LLMs) like GPT-4 and Gemini exemplify the power of deep learning, demonstrating remarkable abilities in pattern recognition, language understanding, and even creative reasoning. Yet, today's most powerful AI systems are not purely connectionist. Instead, they fuse deep learning with symbolic AI, combining pattern recognition with structured reasoning to create more reliable and interpretable models.

The shift from symbolic to connectionist AI was not just a technological shift, it was a paradigm shift in our understanding of intelligence itself. Today, the rise of **hybrid AI models** marks the next stage in this evolution, blending logic with learning, structure with adaptability. AI is no longer just a reflection of Enlightenment rationalism or machine-like precision; it is evolving into something far more fluid and complex.

These developments didn't just shape AI theory; they reshaped its real-world applications, which began with mastering games.

AI's Training Grounds: From Chess to Language Mastery

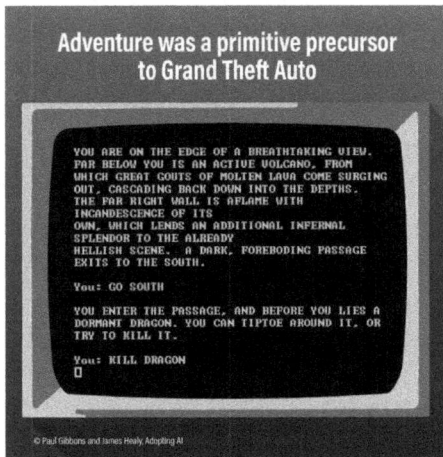

FIGURE II.4: Adventure was a rule-based game with a fixed map and fixed rules, but allowed researchers to experiment with expert systems.

In the 1970s, text adventure games (the one in Figure II.4 was called Adventure) offered worlds where players could select options that led through the game tree. Such games were finite and fixed; the dragon was always in the tunnel. Yet, games such as this contributed to the development of **expert systems**, a milestone on the path to today's AI: computers following strictly scripted routines. Expert systems, while a breakthrough in AI, and still part of the AI landscape, are limited to a narrow domain and to problems where facts and rules can be specified, not, for example, in image and language recognition or in complex dynamic environments.

Most rudimentary chatbots are little better than rule-based approach: they present options (rarely the one you really need), and do little more than connect you to stuff you could have read anyway. Unlike phone bots where you can smash ZERO a few times to talk to a person, chatbots are tougher to circumvent. As in the Adventure game, the dragon was always in the damned tunnel. If you can't get past the dragon, frustration takes the place of customer satisfaction.

When a cardiovascular event happens, there is a lot of clinical data to consider in a very short time (say the patient is clutching his chest): imaging, blood work, medications, observation, EKGs, clinical history, and more. Expert systems can parse that rapidly and make sure that nothing vital is disregarded. Chest pain? Yes. ST-elevation on EKG? No. History of CAD? No. Elevated troponin? Yes. (Note: the real decision tree for chest pain has 50+ nodes, not 4.) The system isn't doing any thinking that hasn't previously been done for it: If X, Then Y, Else Z. This is the hallmark of expert systems, which have been used in medicine since the 1970s. MYCIN, one of the first AI-driven diagnostic tools, followed this same rule-based logic to recommend antibiotics for bacterial infections.

AI becomes a gamer

In the late 1970s, computers began to beat humans at backgammon, and AI developers turned their attention to chess. Gary Kasparov, the then World Champion, initially called his computer opponent,

Deep Blue, an "alien opponent," belittling it and claiming that it was "as intelligent as your alarm clock."

"If-then" decision-tree programming (as in early text adventure games) was insufficient for chess. By the 10th move of any chess game, there are at least 50 million possible board positions; by the 20th move, the number of possible positions is in the quadrillions. Chess programs use a vast database of chess knowledge and algorithms to prune positions. Advances in computing power augmented this "brute force" chess ability. Finally, in 1997, IBM's Deep Blue (the "alarm clock") beat Kasparov. This was a new dawn for AI, beating humans in a game that required complex reasoning. But Deep Blue still wasn't learning and it was still dependent upon humans, relying upon heuristics and variables defined and fine-tuned by chess masters and computer scientists. (Watch this space.)

You might think that Deep Blue would have been the end of chess. Bobby Fisher, a 1970s World Chess Champion, thought computers would ruin chess, turning it into a game of rote memorization. Why play a game that is solved? How wrong. Chess has doubled in popularity since Deep Blue, and AI has transformed the game at every level, particularly at the grandmaster level. Cars are 100 years old, but humans still run, many more of us, and much faster than ever.

What does this tell us about us? Perhaps reflect for yourself on what this might signify for meaning and human flourishing. As AI contributes to art, science, business, and entertainment, will it dumb down or crowd out human thinking? Does it refute the thought that when we live alongside AGIs that we will begin to vegetate in some regards? Or, as in chess, will it push us to new heights of mastery?

Machine learning, deep learning, and neural networks

The limitation of symbolic AI, even after Deep Blue, is that it couldn't learn; it could only follow the rules fed to it. But **machine learning,** that is, training algorithms on large datasets, allows pattern recognition, making predictions, and then refining that knowledge. **Deep learning**, a branch of machine learning, uses **neural networks** that emulate the architecture of the brain to enable even more sophisticated pattern recognition. This led to one of the most exciting (or worrying) developments in AI: it could now teach itself.

Consider something a three-year-old finds trivially easy: recognizing dogs. Classical ML depended on **structured, labeled data**; think of an Excel spreadsheet as a rudimentary example. For early ML to recognize a dog, dog images might be labeled with "fur color," "breed," "eye color," "ear shape," and "size." The AI could be trained on this dataset, but structuring and labeling a dataset is heavy lifting. The world isn't categorized into neat boxes waiting for us to study it.

One example of the challenge was a series of photos of chihuahua faces interspersed with photos of chocolate chip muffins, which became a popular internet meme. (See Figure II.5) A human knows instantly which are dogs and which are a tasty breakfast treat, but trying to articulate rules for deciding which is which is fiendishly difficult.

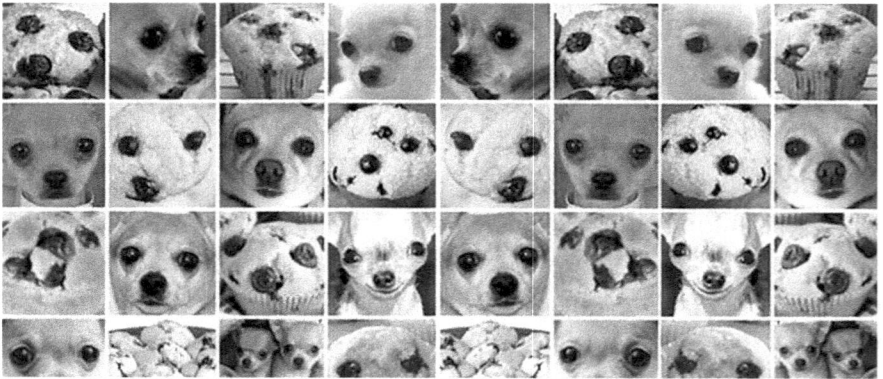

FIGURE II.5: A tasty treat, or a cute companion?

By using **self-supervised learning, clustering, attention mechanisms, and large-scale pre-training** (the P from ChatGPT), AI models can learn to recognize objects like dogs without requiring structured, labeled data. These methods allow AI to detect patterns, relationships, and unique features across vast amounts of unlabeled images, building an understanding of what makes a "dog" image distinct. This approach marks a significant step toward more flexible, human-like learning capabilities in AI, where models can form complex concepts from unstructured information, mimicking some of the ways that humans learn about the world.

Deep Blue could beat human players at chess, but it was close against the best. In 2017, there was another breathtaking development in chess. Deep Mind developed a program called AlphaZero that taught itself chess without any prior knowledge other than what the pieces did. Using another new development in AI, called **reinforcement learning**, AlphaZero played billions of games against itself and in just a few hours could beat the very strongest chess engines, which, in 2017, were already much stronger than human players. Today's chess programs versus humans are like LeBron James playing basketball against elementary school kids. (See Figure II-3.)

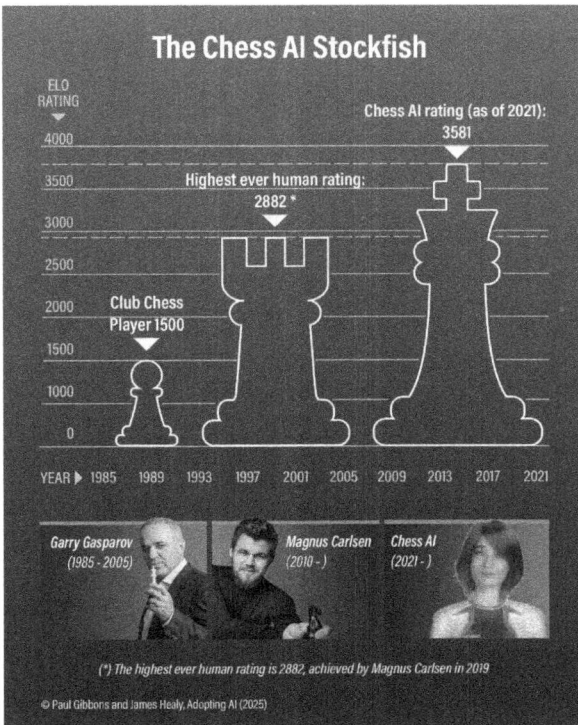

The Chess AI Stockfish

ELO RATING
4000
3500 Highest ever human rating: 2882 *
3000
2500
2000 Club Chess Player 1500
1500
1000
0

Chess AI rating (as of 2021): 3581

YEAR ▶ 1985 1989 1993 1997 2001 2005 2009 2013 2017 2021

Garry Gasparov (1985 - 2005)   Magnus Carlsen (2010 -)   Chess AI (2021 -)

(*) The highest ever human rating is 2882, achieved by Magnus Carlsen in 2019

© Paul Gibbons and James Healy, Adopting AI (2025)

FIGURE II.6: After decades of struggling to beat human players, chess programs have left us in the dust.

The exciting aspect is that computers can now teach themselves complex subjects unsupervised by humans.

What else could they learn this quickly? In which other domains, those more important for humanity than chess, could they excel?

Generative AI, NLP, and transformer architecture

For decades, AI struggled with human language because meaning depends on context, sometimes in adjacent words (*river bank* vs. *investment bank*) and sometimes spread across an entire conversation. Early rule-based systems failed, and even neural networks couldn't fully capture linguistic complexity. Then, in 2017, a breakthrough arrived: **transformer architecture**, introduced in Google's paper *Attention Is All You Need*. By enabling models to process words in parallel and assign importance based on context, transformers revolutionized Natural Language Processing (NLP).

At the heart of transformers is **self-attention**, which allows AI to weigh the significance of words across an entire sentence, paragraph, or document. This means a model can determine that in *the bank of the river*, *bank* refers to land, not finance.

To process text (or any other media), AI **tokenizes** data, breaking it into smaller units and **predicts the next token** in a sequence using probability. If prompted with *The cat sat on the...*, it assigns the highest probability to *mat* based on learned patterns. This predictive ability, powered by massive datasets and computation, is what makes models like ChatGPT so fluent.

Training these systems requires enormous computing power. Models like GPT-4, trained on trillions of words over months, learn from vast datasets encompassing books, web pages, and other proprietary sources. After pretraining, they undergo **fine-tuning** including **Reinforcement Learning with Human Feedback (RLHF)**, where human reviewers refine responses for quality, accuracy, and ethical considerations.

The impact of generative AI extends far beyond text, it powers AI-generated images, videos, music, and code, reshaping industries from entertainment to research. However, while transformer-based AI excels at pattern recognition and content generation, it still lacks true structured reasoning and causal understanding.

This limitation is what the next wave of AI research aims to overcome. **Can AI move beyond pattern prediction and develop real reasoning capabilities?** As we enter 2025 and beyond, the trajectory of AI's evolution, toward Artificial General Intelligence (AGI) or something even greater, remains one of the most critical questions of our time.

CHAPTER

ON THE AI FRONTIER – 2025 AND BEYOND

Chat GPT-4 is the dumbest model any of you will ever have to use again.
Sam Altman

Nothing is worse for writers than discovering that a chapter that they had put to bed months ago has become hopelessly dated. This chapter is like that, "finished" in October 2024, only to be re-written in December 2024 and then again in Spring 2025 (and we venture every Spring for a long time hence).

Deep reasoning models

ChatGPT 3.5 took the world by storm in November 2022, but its successor GPT 4.0, born six months later, was far more impressive.[4] (See Figure III-1.)

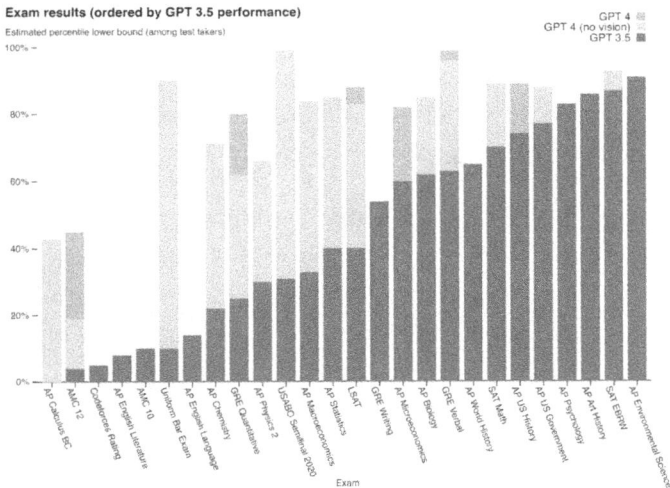

FIGURE III-1: Chat GPT 4.0 (light blue) displayed an incredible array of test-taking skills, from physics to economics to the bar exam leaving its baby brother in the dust. Few, if any, humans could do this. (Source: OpenAI GPT 4 Technical Report)

---

[4] While we've used Open AI's products to illustrate progress in AI (particularly LLMs), there are many competitor models, roughly as impressive, from Anthropic (Claude), Google (Gemini), DeepSeek (DeepSeek R1), Mistral (Le Chat), and others.

Not many humans can perform that well on just one of those tests, let alone quickly master all of them. That kind of progress over just six months (between 3.5 and 4.0) should give us pause for thought: **what will another year bring and another year after that?**

In September 2024, OpenAI unveiled its o1 model, designed to enhance AI's reasoning capabilities. According to Business Insider, this model demonstrated performance comparable to PhD-level expertise in physics, chemistry, and biology. For instance, o1 achieved an 83% success rate on the International Mathematics Olympiad's qualifying exam, a significant improvement over GPT-4o's 13%.

But the world isn't organized into simple, bounded, multiple-choice questions; the interesting and vital questions (drug discovery, materials science, decarbonization) are complex and multi-step. For example, a market entry strategy for a new product might involve market research, economic projections, demographic trends, cultural and regulatory considerations, and planning. Newer LLMs, such as OpenAI's o.1 and o.3 use "chain of thought" reasoning and iteration, matching those that **a team** of expert humans might produce.

Reasoning models approximate structured human thinking through by breaking problems into step-by-step logic (so-called "chain-of-thought" prompting), storing intermediate reasoning steps, and accessing external tools like symbolic logic engines. Some models incorporate reinforcement learning to refine reasoning skills over multiple iterations, while others blend neural networks with symbolic AI for better logical consistency. These improvements enable AI to much more effectively solve math problems, make scientific deductions, and process complex decision-making.

However, breathless headlines aside, the benchmarks commonly used to assess AI performance are not without controversy. Each time AI hits a new personal best, we see far more critical posts than celebratory ones.

There's also a deeper philosophical question here. Imagine sitting down with Einstein, who you've heard is a pretty smart guy. You want to test his intelligence to see for yourself. Which tests would you choose? You give him a Ph.D.-level physics exam, which he mostly smokes, but he gets a few wrong in solid-state physics and quantum gravity, topics which weren't around in his day.

So is he smart, or not? Hard to say. The problem is that tests for intellectual "mortals" don't really help you understand Einsteinian intelligence. AI poses humans the same challenge, new frontier models ace many of the tests we administer to humans. There are some intellectual tasks they still struggle with, but few experts think they will still struggle with these in a year or two.

And as we write this, newer models, more powerful than o1, sometimes called frontier models or multi-step reasoning models, are on the way. At some point, Open AI will release GPT 5.0 Anthropic will release Claude 3.11 Sonnet, and Google Gemini 2.0 Flash Thinking. As of March 2025, the (contested) top four frontier models are shown in Figure III. 2.

A PhD in your pocket?

The newest (frontier) models have generated much excitement among researchers worldwide. Business school professor Ethan Mollick, in his blog, described his experience using DeepResearch (Open AI's hybrid reasoning model) as follows:

> It is, honestly, very good, even if I would have liked a few more sources. It wove together difficult and contradictory concepts, found some novel connections I wouldn't expect, cited only high-quality sources, and was full of accurate quotations. I cannot guarantee everything is correct (though I did not see any errors) but I would have been satisfied to see something like it from a beginning Ph.D. student.

Rough frontier model comparisons March 2025

Model	Key Strengths	Limitations
Gemini Gemini 2.0 Flash (Google)	Advanced multimodal processing, excels in scientific/mathematical reasoning, deep Google ecosystem integration	Limited open access, high computational demand, struggles with nuanced ethical reasoning
✳ Claude 3.5 Sonnet Claude 3.5 Sonnet (Anthropic)	Strong focus on AI safety/alignment, reliable in complex decision-making, interpretable reasoning	Less powerful than GPT-4 in raw performance, conservative outputs, limited creative adaptability
deepseek R1 DeepSeek R1 (DeepSeek)	Competitive with leading models, strong open-source flexibility, rapidly evolving multimodal capabilities	Limited ecosystem integration, geopolitical constraints (e.g., no China-related discourse), fewer independent benchmarks
o3-mini o3-mini (OpenAI)	Surpasses DeepSeek R1 in performance, improved efficiency, reduced costs, and increased speed	Newly released, limited benchmarking data available, long-term reliability not yet established

© Paul Gibbons and James Healy, Adopting AI

FIGURE III.2: Rough frontier model comparisons, March 2025 – is like watching a hotly contested horse race with the lead changing weekly.

That summarizes, more or less, how we feel using it. Paul would have been blown away if one of his first-year graduate students produced a thesis as well researched and argued as the newest LLMs can produce in a few minutes.

But both Paul's and Mollick's evaluations could be called "seat-of-the-pants" and so far there are far too many seat-of-the-pants evaluations, positive and negative, and too few rigorous analyses.

AI does make mistakes and if your authors were using it in fields where we had no prior expertise, we might not catch them. When we wade into alien technical subjects, we opt for peer-reviewed research journals rather than AI summaries.

But to us, today, it feels like having a team of doctoral candidates supporting our work.

The problem is, rigorous peer-reviewed model measurement and evaluation takes time, and there are new, better models being released every week. That means people like you (and us) who aren't AI researchers have to rely on commentary from a few dozen industry experts, discounting some views based on the commercial and career interests most have. And, as social media works, if you are a "doomer" or an AI skeptic, you will read, click, and like views that support your predispositions. And vice versa.

Will scaling continue to improve models?

That is the superheated 64-billion-dollar question in AI research in 2025.

Even the most hardened skeptic will acknowledge that GPT 3.5 was capable of some stunning work. And nobody doubts that the frontier models from Figure III.2 are qualitatively better than that. The debate today is about the future, whether the (unarguable) rate of progress can continue. That debate hovers around the **scaling hypothesis**: that increasing the size of AI models (more parameters, data, and compute) leads to **qualitatively better intelligence,** potentially approaching AGI.

Critics argue that just increasing model size yields diminishing returns, as LLMs rely on tokenization and probabilistic predictions, and lack comprehension. They are, according to the critics, mere "stochastic parrots", lacking embodied experience and causal reasoning; scaling alone won't fix that.

Proponents of scaling counter that **emergent abilities**, unexpected competencies, arise as models grow in size and complexity. For example, GPT-4 demonstrated advanced problem-solving skills, such as solving high-level math problems and improved linguistic nuance in creative writing. These emergent properties are cited as evidence that scaling can lead to qualitative leaps in AI capabilities.

Moreover, leaders in the field, such as Dario Amodei, maintain we haven't seen evidence for diminishing returns to scaling, though critics are guessing it will soon flatline.

Detractors caution that emergent behaviors, while intriguing, do not equate to understanding or consciousness. They argue that without addressing architectural constraints, such as the inability to reason causally or retain long-term memory, scaling alone will not bridge the gap to AGI.

The problem with relying on emergence as a counter to the skeptics is that by definition, emergence is observable but unpredictable, thus speculative. In our view, emergence amounts to a leap of faith, that surprising results, qualitative leaps in intelligence, will continue to surprise.

However, AI researchers counter that scaling is not the only game in town. Hybrid approaches, such as integrating symbolic AI with neural networks, or training methods like reinforcement learning from human feedback, add to the possibility of a qualitatively new intelligence.

The fight goes on. While scaling has unlocked remarkable capabilities, the question of whether it can lead to "real intelligence" or merely more sophisticated pattern recognition remains unresolved.

For today's business users though, this debate is tangential, perhaps irrelevant.

Current model abilities far surpass the requirements to be useful to individuals and businesses; the gap between the technology and its application (outside the research environment) remains huge. If AI research were to stop dead in its tracks tomorrow, it would be many years before use cases catch up.

The rise of the agents

In the workplace, do we want a co-worker that can ace PhD-level physics tests and the Bar exam? People like that can be kind of annoying. No, we'd like AI to do useful things, not just show off its intelligence.

Perhaps you've been experimenting with early LLMs in your work and life, you might have wondered – can they do stuff, not just tell you stuff? The next few years will see a rise in AI agents, that is, semi-autonomous "bots" that can act as well as think for us – or what some call "agentic AI". (See Figure III.3.)

FIGURE III.3: Agents in The Matrix didn't just talk, they could, you know, shoot you.

Agentic AI moves AI from thought to action, answering questions, solving complex multi-step problems, and managing multi-step projects. The complex thinking of more recent LLMs (such as GPT o1) is a advance in thinking power, but AI agents are an advance in **doing power**. There are now thousands of unique AI agents that can solve problems and execute complex processes for us.

To do what it does, an AI agent bridges perception, reasoning, and action, acting as a purposeful entity. But there is a lot of loose use of the word "agent" with some vendors using the word to describe AIs that have very limited functionality. While, technically, the AI that spam filters your email or auto-responds when you are on vacation could be called "an agent", the more interesting ones are far more complex. As more firms trumpet releases of whizz-bang agents, it is worth our readers understanding some of the essential characteristics and differences with other instances of AI. Figure III.4 shows the difference between an **LLM** and an **AI agent**.

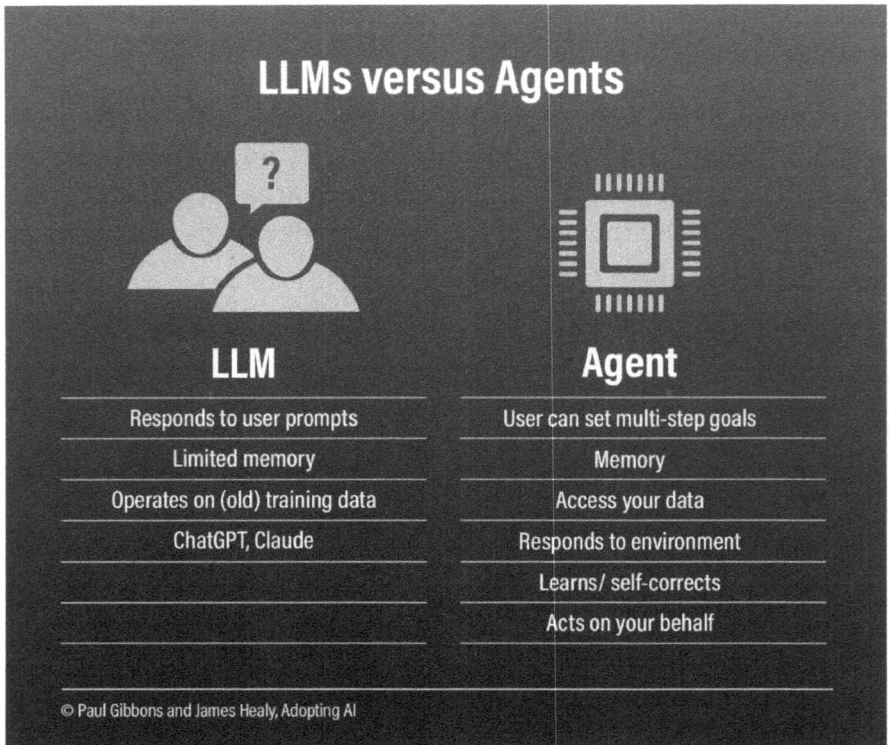

LLMs versus Agents

LLM	Agent
Responds to user prompts	User can set multi-step goals
Limited memory	Memory
Operates on (old) training data	Access your data
ChatGPT, Claude	Responds to environment
	Learns/ self-corrects
	Acts on your behalf

© Paul Gibbons and James Healy, Adopting AI

FIGURE III.4: People use the term agent loosely, but you can think of it as sitting on top of an LLM "directing traffic," stringing together tasks, data, and feedback.

From LLMs to AI Agents

A realtor friend of ours runs a small business and relies on ChatGPT for answering emails, scheduling meetings, sending contracts, and researching leads. However, she needs to manually prompt it for each task, piecing together her workflow step by step. Imagine instead if an AI agent handled the entire process: researching leads, reaching out to the most promising ones, responding to inquiries, scheduling meetings, and finalizing contracts, all with a single push of a button, not through dozens of disconnected prompts.

Agentic AI represents a fundamental shift. Instead of waiting for instructions, agents act independently. Unlike generative AI, which focuses on producing outputs, the most sophisticated agents are goal-driven, sending how to act withing digital environments, autonomously gathering insights, initiating tasks, interacting with systems, and adjusting actions in real time.

For example, while an LLM might draft an email, an AI agent could autonomously detect high-priority emails, generate responses, schedule follow-ups, and track pending replies, all without human micromanagement.

As these systems evolve, they will handle increasingly complex, multi-step tasks, identifying new objectives along the way.

Figure III.5 breaks down the essential vs. optional features of AI agents, distinguishing baseline automation from fully adaptive intelligence.

As AI continues to evolve, look for increasingly complex agents, certainly with more features from the right-hand column and perhaps other features that are still in development.

The seven types of AI agents and their habitats

We hope you will start exploring agents, whether in the workplace, or for booking your next vacation. We encourage you to think about which of the features you recognize from the previous section, and the level of autonomy the agent has. Are there humans in the loop? What proprietary data does it use? How complex are the steps it executes? Does it learn and how?

By the end of 2026 workplaces will be swarming with AI agents. Understanding how they work will help you operate more effectively in a world where agents are increasingly part of the day-to-day. To help with this, Figure III.6 shows seven types of AI agents roughly ranked in order of complexity.

What functional features are essential or optional in an AI Agent?

Essential Features	Optional or Advanced Features
Perception and Input Processing: multi-modal including sensors, data streams, images, voice or text	**Knowledge Management:** use and update domain-specific knowledge and external data (for example, through Retrieval Augmented Generation)
Reasoning: evaluate inputs in the context of goals, objectives, and constraints	**Learning and Adaptation:** use feedback from humans or sensors to improve performance (e.g., reinforcement learning) - essential for dynamic, goal-oriented systems
Acting: including instructing other agents, or interacting with other digital assets	**Memory and Context Management:** the capacity to retain and use contextual information over time, enabling multi-step tasks
	Advanced Actions: using actuators or effectors to undertake complex tasks
	Layered Decision-Making: layering supports more complex projects, perhaps including planning, strategizing, learning, and acting
	Multi-Agent Collaboration: the ability to work collaboratively with other agents
	Autonomous Goal Identification: identify new tasks or goals based on contextual needs – e.g., "Would you like me to generate a sales-call script?"

© Paul Gibbons and James Healy, Adopting AI

FIGURE III.5: What capabilities does an AI agent have to have, essentially and optionally?

Reflex Agents are the most basic, responding instantly to specific conditions without memory or learning.

Email **spam filters** flag or block messages based on "if X, then Y" rules.

Model-Based Reflex Agents go a step further by maintaining an internal model of their environment.

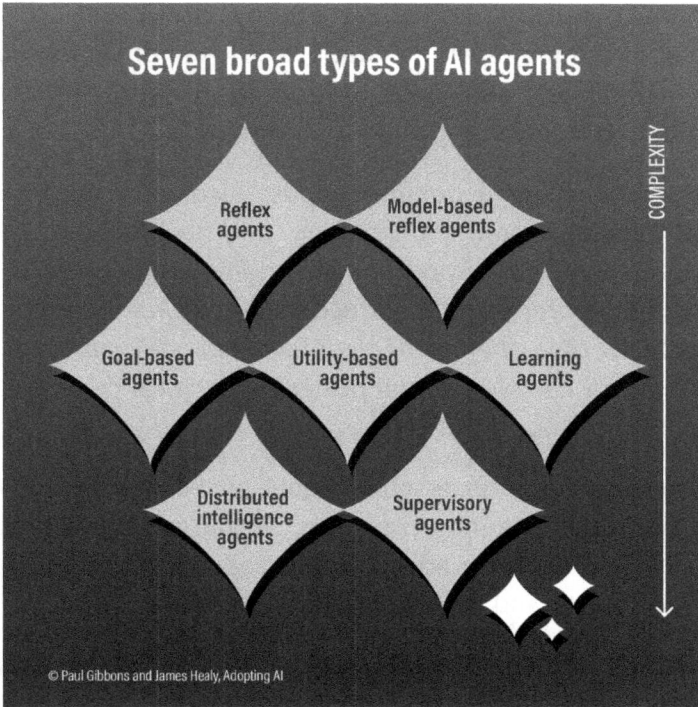

FIGURE III.6: Seven types of AI agents ranked in rough order of complexity.

The **legal** profession has long been labor-intensive, requiring hours of research, drafting, and strategizing. An agent can analyze case files, extract relevant facts, and identify potential arguments or counterarguments while another can review precedent cases and identify key rulings or legal trends to inform strategy.

Goal-based agents operate on goals, not tasks, and can sequence tasks to achieve goals.

Marketing campaigns often require managing a labyrinth of tools and processes, from ideation to execution. Marketers can simply describe their campaign goals and target audience in natural language, and agent systems translate these inputs into actionable strategies.

Learning agents continuously improve by analyzing past data and refining recommendations. They might track workflows, timelines, and collaboration to optimize efficiency, predict bottlenecks, and suggest training.

__Healthcare__ learning agents support medical professionals in diagnostics, treatment planning, and patient care. These agents analyze patient data, medical imaging, and research findings to improve diagnostic accuracy and personalize treatments. Over time, they refine their recommendations based on new medical knowledge and patient outcomes, continuously enhancing the quality of care.

Utility-based agents resemble goal-based agents but can weigh and balance multiple goals. Resource allocation systems can dynamically adjust workloads across teams, reassign personnel, and optimize budget spending.

One major retailer uses a __logistics__ system that can reallocate staff, adjust inventory distribution, and optimize budget spending across multiple store locations, balancing multiple objectives, such as cost efficiency, workforce allocation, and customer satisfaction.

Supervisory agents are an instance of a **multi-agent system**, an agent that can oversee other agents. They can take the place of human supervisors chaining agents together, further automating agentic workflows. To do so, they combine features of less complex goal-based, learning, and utility agents and are still a rare species.

In a complex logistics operation, a supervisory agent oversees a fleet of autonomous delivery drones (goal-based agents) and inventory management bots (utility-based agents) within a regional warehouse network. The supervisory agent dynamically allocates tasks based on shifting constraints—weather, delivery urgency, traffic, and stock levels—while monitoring performance and retraining the underlying agents when inefficiencies arise.

Finally, **distributed intelligence agents** operate in multi-agent systems but with decentralized decision-making (unlike a supervisory agent) and more human-like collaborative behavior. These agents operate autonomously but communicate and coordinate their actions, enabling efficient, scalable solutions in dynamic environments. They are, as of this writing, aspirational because to collaborate as described, agents require autonomy. We, however, expect to see concrete instances within a year. Winding that distributed intelligence agent story forward, we can imagine an entire small enterprise, an **AI-only company,** at some stage in the near future.

From writing and marketing to healthcare and supply chains, AI agents are revolutionizing how work is done, unlocking new levels of efficiency, creativity, and value. As these systems mature, their integration into the workforce will only deepen, creating an ecosystem where humans and AI collaborate to achieve shared goals.

AI 2030: powering AI

> "Pray, gentle guide, shape well this newborn power,
> Lest in its wake all realms of man devour."
> (By GPT o1, during a Guardian test of its Shakespearean proficiency)

As we journey toward 2030, the AI landscape will be defined by more than just frontier models and advanced agents, it will be shaped by innovations in hardware, infrastructure, and perhaps even **quantum computing**. Breakthroughs in **specialized AI hardware**, like **neuromorphic processors** and **photonic computing**, promise leaps in efficiency. Improvements in **accelerator chips and specialized inference hardware** are driving down the costs of inference significantly, enabling AI labs to offer models at lower prices.

That word salad means cheaper, faster, better processing. For example, the cost per million tokens of OpenAI's top models has dropped by over 200-fold in the past two years. The good thing about that is Sam Altman will make more money. Just kidding, what it means is a democratization of AI. Today, the bare minimum to train a proprietary LLM is low seven figures, and to run it (inference) costs low six figures annually. That should enable disruptive startups, which would be good for us all.

As AI becomes more affordable, we must ask: Will simply scaling compute and data continue to drive breakthroughs, or are we approaching a plateau? The next few years will test the scaling hypothesis that more compute and bigger data sets produce qualitative leaps in intelligence. Should AI inference and accuracy flatline, that will challenge researchers to develop new architectures. The best models now combine symbolic and connectionist approaches. But alternatives like **state-space models (SSMs)** and **liquid neural networks (LNNs)** are gaining traction. These architectures tackle the memory and compute bottlenecks of transformers, achieving **linear complexity**, where computational cost grows proportionally with input size. (Trust us, this is much better than quadratic or exponential scaling, which explodes computational demands.)

Einstein once (perhaps apocryphally) said, "If I had an hour to solve a problem, I would spend 55 minutes thinking about the problem and five minutes thinking about solutions." His insight underscores a crucial point: deep reasoning matters more than quick answers.

Large reasoning models (LRMs) represent a new frontier in problem-solving. A large language model (LLM) can generate a PhD-level analysis in mere seconds, **but what happens when it thinks for an hour, a day, or even longer?** The challenge isn't just speed; it's structuring a chain of thought, breaking problems into steps, and iterating between problem space and solution space.

To harness AI's potential, we need to develop **AI meta-cognition**: the ability to direct and refine its own reasoning, much like a skilled thinker does. That's the key to unlocking transformative AI.

Nobel Prize winning behavioural economist Daniel Kahneman once said: "prophecy is for fools". We don't know where AI is going, but we are certain it's going to play a bigger and more influential role in almost all aspects of human affairs. Adopting, and adapting to, AI is arguably the single most important strategic imperative for organizations in the 2020s. It's to that adoption that we turn next. What is on the applied AI frontier?

AI 2030: the adoption bottleneck

Today's large language models far exceed what's needed for most business applications. More powerful AI is needed for science, to unlock breakthroughs in materials science, drug discovery, and climate modeling. But many enterprise use cases, customer service, automation, and data analytics, are now relatively trivial for AI. The real bottleneck is no longer intelligence. It is adoption, integration, and human strategy.

The real challenge isn't whether AI is smart enough, it's whether organizations are strategic enough to wield it effectively. As Dario Amodei puts it, the focus should be on *"return on intelligence"*; not just acquiring powerful AI, but ensuring it delivers value. Intelligence alone isn't the limiting factor; organizational design, adaptability, and decision-making structures are.

We've built the tiger, but riding it requires intent, discipline, and direction. This demands a new way of thinking about AI adoption: not just in terms of efficiency gains, but as a driver of competitive strategy, governance, and sustainability. How do organizations prepare for a world where AI plans and executes, not just assists? How do they balance the efficiencies that come from autonomy with oversight? And how do they ensure their AI strategies evolve as fast as the technology itself?

We now turn to AI strategy: how organizations can move from experimentation to full-scale adoption, embedding AI into their DNA, and turning it into a true competitive advantage.

SECTION III

PEOPLE-FIRST AI STRATEGY, ADOPTION, AND WORKFORCE

CHAPTER **IV**

PEOPLE-FIRST AI STRATEGY

If I had asked people what they wanted, they would have said faster horses.
usually attributed to Henry Ford

n early 2024, The Economist reviewed AI adoption rates:

"Most firms do not currently use ChatGPT, Google's Gemini, Microsoft's Copilot or other such tools in a systematic way, even if individual employees play around with them… only about 5% of American firms of all sizes said they used AI. A further 7% of firms plan to adopt it within six months… the numbers conceal large differences between sectors: 17% of firms in the information industry, which includes technology and media, say they use it to make products, compared with 3% of manufacturers and 5% of health-care companies."

Whom to believe? Surveys from consulting firms place this much higher – nearly two-thirds of enterprises are using AI, according to McKinsey. The difference is explained (partly) by a difference between Gen AI and earlier narrow-AI deployments. McKinsey's method asked about adoption in any single function – perhaps legacy use cases in cybersecurity or chatbots would count. It seems to us that perceived social norms may be a factor – firms are more likely to exaggerate their efforts (because Wall Street) than downplay them, and McKinsey have a vested interest in creating FOMO among potential clients. Anecdotally, in our travels among hundreds of mid-sized companies, The Economist's data are closer to reality – we met very few senior executives who were much past the AI-curious stage.

Amid the proliferation of hype (not least in this book's first chapter), you might wonder, why is AI adoption by businesses proceeding at such a sluggish rate?

We are reminded of internet adoption. The world wide web launched in 1993, but it took almost two decades to reach the tipping point that two-thirds of businesses in the US had a website. And having a website is a long way from digital transformation. Many firms simply took their marketing material, digitized it and stuck it on the web as "brochureware". It wasn't until the 2010s, with the advent of smartphones, that businesses began to realize they were leaving money on the table. The transformative effect on work and workplaces was arguably only felt nearly 30 years later during the global COVID-19 pandemic as stay-at-home mandates forced businesses across a range of sectors to radically virtualize their operating models and finally prioritize digital first.

While most established corporations just tinkered with the Internet for two decades, the so-called "FANGs" (Facebook, Amazon, Netflix, Google) ate their lunch. Bricks-and-mortar book sales fell 75%, and Amazon captured 50% of the $20bn book market. PayPal birthed the online payments industry; Facebook crushed the social media businesses of Friendster (huh?) and Myspace (who?). Netflix invented an entirely new category, wiping out a range of established competitors up and down the media value chain.

Sure, in early 2025, the hype herd may be stampeding toward an AI-enabled future, but most executives will remember management fads that generated more heat than light: Business Process Redesign, Six Sigma, Total Quality Management, the Metaverse, and the Y2K "bug". Silver bullets have an uncanny tendency to miss the mark. Many will remember the 90% crash in the NASDAQ during the "dotbomb" crisis at the turn of the century, when the "next big thing" had an uncomfortable collision with reality. Are we at max hype, or internet-1993? Is the party just getting started?

Kodak versus Netflix: hubris versus humility?

Two established 1990s businesses illustrate the effect new technologies can have and hint what might be required to adopt AI. Kodak invented digital photography, launched the first digital camera in 1991, and had roughly a 50% share of the US photography market across all product lines. **That is about as good a head start as can be imagined:** the first mover in a new technology with unmatchable access to customers and capital.

Yet, they declared bankruptcy in 2012. How this happened is etched into business history as a lesson in what **not** to do with transformational technology. Their cash cow was film and printing, and digital photography threatened to cannibalize that. Their mindset was, "We are a chemical/film company and not a technology company." They made atrocious strategic decisions, thought time was on their side, continued to focus on print-centric digital products, and under-invested in the revolutionary new technology they themselves had developed.

Mindset, culture, and strategy are a reasonable three-word summary of what happened, but given mindset and culture determine strategy, the first two merit particular attention. Senior leaders of today would do well to review Kodak's history with some questions in mind: "Could this happen to us?" "What does our current culture imply about adopting AI?" "What assumptions are we making?"

Netflix is the classic counterexample. Like Kodak, they had a substantial market share, 20 million subscribers and 90% of the DVD-by-mail market: a very comfortable business that could have fallen prey to a complacent mindset. Somehow, Reed Hastings and his team were courageous enough to see that their business model was dead, unlike their competitor, Blockbuster, who watched Rome burn. Somehow, they had the courage and foresight to invest in streaming infrastructure before there was a streaming market.

Next, they added content development to their strategy, riskily taking on giants such as Disney, Warner, Paramount, Sony, NBC, etc., despite a profitable and comfortable position streaming content leased from those giants. They scaled technology infrastructure to serve 200m subscribers in 190 countries and forever altered cultural norms with their binge-watching release model. "Watching Netflix" is how people describe watching TV, even when watching Apple, Hulu, Prime, or another service (in the same way that Xeroxing became synonymous with photocopying).

Netflix, repeatedly, were willing to cannibalize their existing businesses, to view disruptive technologies as opportunities, rather than hunkering down to protect themselves from threats, and have a legendary business culture (e.g., no rules *rules*, radical candor, and freedom with responsibility).

The question for today's AI age is whether non-native AI firms will be Kodak or Netflix? One way of pursuing that is to view strategy as a human undertaking, inseparable from our social, emotional, tribal, and storytelling nature.

Animal spirits and strategy

"Most of our decisions to do something positive, the full consequences of which will be drawn out over many days to come, can only be taken as the result of animal spirits—a spontaneous urge to action rather than inaction, and not as the outcome of a weighted average of quantified benefits multiplied by quantified probabilities."
(J.M. Keynes)

Like much human decision-making, business strategy is often idealized as an exercise in rationality: ROI, benchmarking, mathematical modeling, financial forecasting, and so on. That is neither descriptively true nor ideal. Keynes' evocative phrase, "animal spirits", conjures up a vast academic literature spanning anthropology, behavioral economics, evolutionary biology, neuroscience, psychology, and sociology that shows we aren't analytical, calculating machines.

The rational strategy model would not have predicted the divergence in outcomes between Netflix and Kodak. In caricature, one firm proceeded with humility (about the sustainability and robustness of their business model) and courage. The other seemed ego-driven, dining out on their market position (#1), their inventor status (first) and their identity (as a chemical/film company.)

Viewing strategy as a purely rational exercise is not even an ideal idealization. **(Not only does it not work that way, it shouldn't.)** During implementation, "animal spirits" such as passion, resolve, and grit matter a great deal. Our lives are full of strategies we should rationally pursue but fail to, in say, lifestyle modifications or weight loss. We've seen senior leaders aplenty craft an analytical strategy without asking the asking themselves whether anyone is passionate about pursuing it.

Consider wholesale organizational transformation, which might take several years and include several wrong turns and setbacks. Animal spirits will matter even more during the tough times.

AI attracts human responses, from excitement to anxiety, from confidence to hesitation, from hubris to fatalism, from hate to adoration, and so on. The humans you work with on a new use case, simple or complex, may not trust or understand AI, even at senior level, where they ought to by now, but often don't.

A people-first approach to strategy recognizes humans as we really are, not as management gurus wish us to be. Teams must grapple with the animal spirits, whether excitement or anxiety, confront their assumptions, and honestly examine their level of commitment and competence before "rationally" choosing a strategy.

In the next chapter, we continue the people-first focus with a closer look at AI adoption with an emphasis on culture. In the final chapter of this section, we look at people-first workforce strategies. First, though, let us consider two shifts that organizations face when considering their AI strategy.

Don't think technology, think intelligence

As the AI hype train gathers speed, your authors have encountered more than one concerned executive breathlessly asking, "what's our AI technology strategy?" Arguably, the biggest pitfall for organizations considering their technology strategy is **considering AI a technology, not an intelligence.**

Considering AI as an intelligence reflects the transformative potential of AI and helps understand its myriad ethical challenges. This pivots the strategy question from "which problems can this new technology help solve?", to considering "how could more intelligence help us?" Shifting from thinking of AI as merely a technology to seeing it as an evolving intelligence has implications for business strategy, leadership, and innovation.

Viewed as an **intelligence**, AI has much greater potential becoming a dynamic learning entity that augments and sometimes surpasses human cognition. That allows organizations to start to rethink roles, structures, and decision-making at a deeper level. But AI can do more than just automating workflows. Thinking intelligence, not technology, allows leaders to driver business model reinvention, reshaping how value is created, delivered, and monetized.

The most transformative implication is that organizations embracing AI as intelligence rather than a tool are more likely to build businesses that are adaptive, resilient, and self-improving. They won't just integrate AI into operations, they'll restructure around it, making AI a co-pilot in strategy, innovation, and decision-making. This shift demands a brave new kind of leadership, one that treats AI as a partner in shaping the future.

Don't think use cases, rethink business models and processes

While it is natural to focus on automation, efficiency, and incremental improvements to optimize existing processes, this leads to a **tactical "use-case mindset,"** where AI is adopted piecemeal, missing strategic opportunities. Yes, augmenting existing processes, such as automating customer support, optimizing supply chains, or enhancing marketing analytics, are valuable, but may miss the opportunities that come from being strategic and not tactical.

This happened frequently in digital transformation. There were **digital tourists,** who added disconnected digital capabilities here and there, but missed the transformation big picture. By rethinking business models and processes first, businesses can avoid the "technology tail wagging the business strategy dog."

Kodak vs. Netflix is again instructive. Kodak, despite inventing the digital camera, remained locked in a use-case mindset, using technology to improve film-based photography rather than redefining its business model. By the time digital photography disrupted its core revenue streams, Kodak could not adapt. Netflix, on the other hand, ran toward transformation, shifting from DVD rentals to streaming and to AI-driven content recommendation and production. Rather than using technology simply to refine DVD delivery, Netflix integrated technology into its entire value chain, reshaping entertainment consumption and content creation itself.

The contrast between Ford and Tesla illustrates this today. Ford largely follows a use-case AI strategy, applying AI in incremental ways, enhancing supply chain logistics, improving driver assistance systems, and optimizing manufacturing. While AI plays a role, the company still operates as a traditional automaker. Tesla, however, represents the transformational approach, using AI not just to enhance vehicle performance but to redefine the very nature of mobility, pioneering autonomous driving, AI-powered smart energy solutions, and even robot manufacturing.

All firms, wherever they are in AI deployment, must decide how fast to go, and whether to deploy AI faster than the competition, or wait until prices come down, bugs are ironed out, and lessons have been learned.

Lead, match, or follow?

"Not to decide is to decide".
(Harvey Cox, American theologian)

The decision faced by senior executives and business unit and functional heads is whether to lead, match, or follow. This doesn't apply just at the macro, corporate level. A function head or business unit leader will face similar strategic decisions; a CHRO, for example, may need to decide whether to invest in a newly released HR agent.

Like all such broad choices, several factors matter: competitive position, risk tolerance, and market dynamics, among many others.

As Reid Hoffman, former CEO of LinkedIn, frames it, "Businesspeople have to understand that speed of innovation is not just set by internal rumblings, their stomachs, or in-house meetings, but by the industry as a whole - and by what their competitors, suppliers, partners, distributors, and society are doing."

Too quick or too slow? Death by AI

While organizational AI adoption might not yet have reached a critical mass, AI disruption has already torpedoed some companies. Chegg was an ed-tech giant that pioneered textbook rental and then expanded into homework, tutoring, and test prep, giving it a billion-dollar business. ChatGPT upended Chegg's business model, which had blossomed during the pandemic years, offering students better and more tailored support for free or a fraction of the cost. Chegg's share price fell 99% over four years, and while they're exploring AI integration, revenues are still tumbling and it all feels too little, too late.

In contrast, digital health provider Babylon Health was quick off the mark. Founded in 2013, Babylon was quick to embrace AI-powered telehealth and diagnosis solutions. They were too quick. The technology was not quite robust enough and their diagnoses were sometimes unsafe or inaccurate. Moreover, their claim that their technology outperformed human doctors was, at best, misleading. They listed with a 4-billion-dollar valuation in 2021 but went bankrupt in 2023 and were eventually sold for under a million dollars.

Many of Babylon's challenges were ethical. They were a well-capitalized, early adopter with a major client in the UK's National Health Service. However, operating in a highly regulated health environment requires extreme attention to compliance, robust clinical trials, and an evidence-based approach. They struggled with algorithmic bias, and the lack of transparency in their model's decision-making process eroded client and regulator trust. Their enthusiasm got the better of them and they made incautious claims about what their tech could do. The final section of the book contains a lengthy treatment of how these types of issues ought to inform every organization's approach to adopting AI.

Babylon Health's story is both a cautionary tale and a learning opportunity. While its vision was transformative, its execution exposed the pitfalls of allowing AI hype to outpace transparency and trust. We don't pretend this dilemma has an easy resolution. On the one hand, firms can't afford to be overtaken, to look stupid as competitors eat their lunch, or watch feisty startups grab scale. On the other hand, adopting AI too hastily may result in poor choices with poor value realization.

These case studies illustrate the perils of moving too quickly or too slowly. The path to a "Goldilocks" solution to that dilemma is linked to the crucial strategic question of whether to be a leader or follower.

Following

Following looks less risky, particularly for highly cost-constrained companies or those in industries where AI appears to have minimal impact. In later stages of development, turnkey solutions may make implementation easier. As technology matures, costs tend to come down, so waiting might be prudent economically. Waiting also allows ethical and regulatory risks to be evaluated, reducing investment risk.

Leaders also need to consider the market dynamics of their industry. In some sectors, including creative, craft, mental health, caregiving, luxury brands and experiences, and early childhood education, the pace of disruption is likely to be slow because of their innate "humanness." (Finance, technology, and technology-services are the opposite.) See Figure IV.1 for an overview of industry adoption rates.

Following, though, should be a conscious choice, not a lazy or complacent one, nor one grounded in ignorance or denial. The theologian's mantra, above, is a reasonable guide: failure to decide is very much a decision.

Industries that appear to be moving fastest and slowest in AI adoption

Industries Adopting AI Fastest/Most	Industries Adopting AI Slowest/Least
Financial Services (Banking, Trading, Insurance)	**Skilled Trades** (Plumbing, Carpentry, HVAC)
Healthcare (Diagnostics, Personalized Medicine, Admin Automation)	**Artisanal & Craft Industries** (Handmade Goods, Fine Arts)
Transportation & Logistics (Autonomous Vehicles, Supply Chain Optimization)	**Spiritual and Caregiving Services** (Clergy, Care Industries)

© Paul Gibbons and James Healy, Adopting AI (2025)

FIGURE IV.1: AI industry adoption, leaders and laggards

Should leaders **consciously** choose this stance, they must find another way to compete with firms moving more quickly on AI. They might use the resources that might have been invested in a leading or matching strategy to differentiate themselves in other ways, perhaps even in contrarian ways emphasizing the human touch aspect of your brand.

Matching

Matching is sometimes called a "fast-follower" strategy. It involves adopting AI quickly after pioneers, but improving on their mistakes or inefficiencies. Evaluating competitor adoption is key. If competitors are successfully implementing AI, it could signal an opportunity to improve upon their own business models. Cost vs. ROI optimization also plays a critical role: can you adopt AI more cost-effectively than first movers, using mature, off-the-shelf solutions that minimize investment risks?

Customer expectations and technology readiness further influence the decision. If AI has shifted from a differentiator to a hygiene factor in your industry, delaying adoption could risk brand relevance and market share. This requires assessing whether current AI tools are stable and accessible, and whether your IT infrastructure can support seamless integration without significant overhauls.

Matching may be most appropriate in mid-sized organizations that can't afford the risk of a first-mover strategy, but who, because of their size, may be ripe for disruption if they ponder for too long.

Leading

In contrast, **leading** takes a leaf from the Netflix playbook: invest, pioneer, and innovate and set a pace for competitors to follow. However, as always, context is key, not least in terms of industry and market maturity. Is AI transformational within your sector, or is it still in an experimental phase? If competitors are slow adopters, this presents an opportunity to gain a competitive edge.

Leading strategically will depend on several questions:

1. What pace of AI deployment can our talent and infrastructure handle?
2. Do we have access to advanced AI R&D through partnerships or suppliers?
3. Can we meet compliance challenges and regulatory scrutiny?
4. Do we have unique, proprietary data that will fuel our differentiation? (Using the same data as everyone else won't yield any advantage.)

AI strategy is not, as the next few chapters underscore, just a technology strategy, but neither can it be divorced from technology strategy. Fast or slow, lead, match or follow; these strategic dilemmas are inseparable from the question familiar to every CTO: build or buy?

Build, buy, or ecosystem?

"I am tomorrow, or some future day, what I establish today. I am today what I established yesterday or some previous day."
(James Joyce)

When considering their strategic choices around AI, organizations need to recognize the constraints created by the nature of AI and the (for most) prohibitive cost and complexity of building native AI, such as proprietary LLMs. Most strategies likely involve at least some buying from one of the major AI vendors and/or an enterprise software provider embedding AI into their products.

Build and buy strategies

Both your authors have encountered senior leaders convinced that their organization's exciting new AI platform is a bespoke model their tech team built and trained in-house. That is almost never reality given the capital required to train even a modest one.

OpenAI and Google offer sophisticated enterprise packages providing access to ring-fenced instances of their frontier models on which organizations can build customized applications and integrations with their existing tech stack.

Building new solutions offers control, customization, and competitive advantages, particularly when proprietary data, specialized fine-tuning, or bespoke workflows are involved. Building also lends itself to far more transformative approaches by allowing organizations to rethink and restructure products, processes, roles, and organizational structures. However, building necessarily requires significant and ongoing investment in AI talent and infrastructure. It also increases the risk of failed adoption if the necessary changes to value chains, processes, and structures aren't expertly designed and implemented.

Given the level of internal capabilities required, talent and infrastructure, most medium and smaller businesses will buy.

Buying accelerates deployment, reduces costs, and reduces the adoption risk for organizations because the vendor has already figured out how to integrate AI into their existing process or workflow.

The downside is that vendor-provided standard functionality is, well, standard; if you have it, chances are so do all your competitors. Buying locks an organization into the incremental use case approach to AI deployment outlined earlier. There are also potential data privacy concerns, dependency on external providers, vendor lock-in, and the risk that inattention to developing internal AI capability will leave an organization dangerously exposed further down the track.

Often choosing between build versus buy will be a choice between **proactively** creating bespoke new solutions underpinned by AI models bought (or more accurately, "rented") from major AI vendors, or, **reactively** adopting the AI features added to existing enterprise applications by major software vendors like SAP, Oracle, Salesforce, Microsoft, or Atlassian.

In practice, most organizations will do a little of both. This is more complex than buying off-the-shelf AI infrastructure and agents but requires less capability than build-only strategies.

Ecosystem (hybrid) strategies

Most IT departments have traveled this road many times with enterprise software and the public versus private cloud debate. Much as hybrid cloud deployments offer a nuanced compromise, there's much to be said for an **AI ecosystem strategy**, blending in-house development with partnerships across enterprise technology providers, cloud platforms, startups, and academic institutions.

This hybrid model enables companies to build AI capabilities while benefiting from external expertise and infrastructure for non-differentiating components. For example, a financial services firm might develop proprietary AI models for fraud detection while integrating third-party AI solutions for customer service automation. This approach reduces reliance on a single vendor, enhances flexibility, and ensures that AI investments align with both short-term and long-term strategic goals.

Two examples of the ecosystem approach come from AI-native Nvidia, and very much non-AI-native GE Healthcare.

Nvidia has built an AI ecosystem of hardware, software, and partnerships that supplies GPUs and AI accelerators for deep learning. In parallel, it has developed platforms like CUDA, TensorRT, and Omniverse to support AI developers. Nvidia also collaborates with cloud providers, research institutions, and enterprise AI adopters to drive innovation across healthcare, autonomous driving, and generative AI. This ecosystem approach has solidified Nvidia as much more than a hardware supplier.

GE Healthcare's ecosystem approach partners with hospitals, research institutions, and tech firms to drive AI-powered medical imaging and diagnostics. Its Edison AI platform enables hospitals and developers to integrate machine learning models into radiology workflows, helping doctors detect diseases faster and with greater accuracy. Instead of developing all AI solutions internally, GE Healthcare collaborates with startups, academic researchers, and cloud providers, ensuring flexibility, scalability, and rapid innovation without overextending its in-house capabilities.

The ecosystem approach to AI adoption accelerates innovation by fostering collaboration and reducing time-to-market. Companies can tap into open-source AI frameworks, cloud-based AI services, and industry-specific consortia to stay ahead without bearing the full burden of AI development. By co-developing solutions with partners or participating in AI-driven alliances, businesses can reduce risk, share knowledge, and scale AI adoption efficiently. This strategy is particularly useful for industries where AI is rapidly evolving, ensuring adaptability without locking into rigid buy-or-build decisions.

While the build versus buy debate echoes earlier non-AI technology strategy dilemmas, we reiterate the crucial point that this should not be a purely tech-centric strategy. Nvidia CEO Jensen Huang made headlines in January 2025 when he claimed, "In a lot of ways, the IT department of every company is going to be the HR department of AI agents in the future."

Your authors are firmly of the view that humans are far too important to be solely the responsibility of the HR department; similarly, AIs are far too important to be solely the responsibility of the IT department.

Strategy to Adoption:
Untangling the threads

Any sound strategy development process recognizes the uneasy interplay and trade-offs between different facets of strategy. AI will affect almost every part of an organization in some way, making the "AI strategy" a tangled web of disparate strategic perspectives: risk, data, people, brand, customers, finance, and technology.

The dizzying rate of AI improvement means assumptions in each of those areas change, constraints disappear, new strategic threats and opportunities arise overnight, making flexibility and adaptability key. As we'll see in the next chapter, that culture of flexibility, adaptability, and innovation is not just crucial to defining the AI strategy, it's also the key to effective AI adoption.

CHAPTER V

ADAPTIVE ADOPTION

"We cannot solve our problems with the same thinking we used
when we created them."
(Albert Einstein)

Culture and AI adoption

Innovations take root in organizations in a cultural soil: too acid, too alkali; lazy or overzealous gardeners; competition with roots from other species and so on. This has been true throughout the history of technology, but with AI, there's a difference. This chapter looks at that cultural soil, why traditional change management won't work, and why adaptive adoption will work better.

Everybody sort of knows what a CRM system is supposed to do and very few fear that a CRM system will steal their job. No-one, we assume, worries that Salesforce will go rogue and wipe out humanity.

But AI is different: it arrives wrapped in hype, fear, and misunderstanding. Only half of respondents in a global survey think the benefits outweigh the risks, with Asian countries the most positive, European countries the most negative, and Americans somewhere in the middle. Over one-third of respondents think their job is likely to be eliminated by AI, and almost four-fifths want governments to step in and deliver AI-specific regulations to ensure it's used responsibly.

Ignorance is rife. When presented with six simple AI use cases, only 30% of business people could identify them all. A full 57% of executives feel their leaders have insufficient knowledge of AI[5]. Even social media for business professionals, LinkedIn and Medium, for example, seem to have as many horror stories as hype, as many trash talkers as tech triumphalists.

A cultural soil that is simultaneously over-fertilized with hype and poisoned with fear makes AI adoption less about technical capability and more about perception, trust, and willingness to engage.

---

[5]   From Pew Research, the World Economic Forum and Boston Consulting Group.

While an agent that summarizes meetings or replies to emails is both simple and understandable, to prosper in the intelligence transition organizations will have to go far beyond such conveniences and undertake more ambitious deployments.

We asked 100 workshop participants to use ChatGPT to create a business and marketing plan. It was illustrative to watch their initial struggles: "How do I install it?" "What do I ask it?" But as time in this learning environment went on, they began to ask it to "research this demographic" and "estimate the price elasticity" and "which suppliers look most attractive" and so on.

As the user gets better, the technology gets better.

As participants became more comfortable, their questions evolved from simple data retrieval to more complex problem-solving. One technology architect eventually asked the AI to design a simulation that would generate code, stress test, and optimize their system using predictive analytics and anomaly detection. But unless prompted to do such things in a learning environment, it might not occur to workers, during the stress of the working day, to do so.

AI is not just another technology, it is an intelligence that learns, adapts, and makes decisions in ways that **blur the line between automation and cognition**. This distinction matters because it shifts the focus from *installation* to *integration*: from "how do we plug this in?" to "how do we work with it?" Successful AI adoption isn't about forcing technology into existing workflows but about reimagining workflows around a new form of intelligence.

This is complicated by the "adoption gap", the yawning chasm between AI's capabilities and our ability to integrate it meaningfully into organizations and society. Dario Amodei, CEO of Anthropic, has suggested that AI development is far outpacing our ability to govern, regulate, and absorb its impact. We said earlier that even if AI research stopped tomorrow, we would still need years for implementation to catch up.

Part of this is a matter of speed: AI operates on a scale and time-line that human cognitive processing simply can't match. But it's not just about human speed. Where AI moves at the "speed of sil-icon", organizational systems, policies, and regulatory frameworks respond far more sluggishly. The result? An AI ecosystem where ca-pability outstrips capacity, leading to a chaotic landscape of partial implementations, abandoned pilots, and ethical blind spots.

AI that isn't understood, trusted, or integrated with human ex-pertise will fail not because it doesn't work, but because it doesn't fit.

Overcoming this logjam requires a people-first approach to adoption.

Millions of words have been spilled debating the extent to which AIs resemble humans. Questions of AI consciousness (whatever that even means), AI explainability, and AI bias are a subset of a much larg-er research effort to try to understand these strange new intelligences we've created. One research paper from Google's DeepMind team, en-titled *Machine Psychology*, suggests the workings of AI are so opaque and impenetrable that techniques from human psychology should be applied to reverse engineer insights from the observed behavior of AI.

These efforts are both fascinating and crucial, as we explore fur-ther in the chapters on AI ethics. But machine psychology is only half the story. Unless we also properly understand human psychology, we have no hope of effectively navigating the collision of human and machine. In other words, culture, behavior, and organizational de-sign are just as critical as model architecture and training data.

Most business publications, from CIO magazine to HBR, and con-sulting firms, from McKinsey to Accenture, frame the solution to this as change management. When we've talked to business leaders about the process for adopting AI, many have said, "Oh, you mean change management?"

No, we don't. It would be a big mistake to use traditional change management for AI adoption.

Traditional change management misunderstands humans

We offer this critique of change management as insiders, with more than a combined 40 years in the change profession. Despite, or perhaps because of, that insider status, we're among the profession's fiercest critics and are particularly worried that traditional change management will be the go-to method for AI adoption. That seems already to be the case, and much of what we read about AI and change seems to be little more than the same old content, updated with a "find and replace" swapping "AI" for "SAP" (or "Salesforce", or "Cloud", or whatever).

Both your authors have written at length about the flaws in traditional change management. Together with Tricia Kennedy and a talented group of collaborators, we co-wrote *The Future of Change Management* in 2024. Paul's *The Science of Organizational Change* is frequently in the top books on the subject on Amazon and has been called one of the best five change management books of all time.

While we've both encountered some deep thinkers in the global change management community who appreciate ideas such as emergent change, systems thinking, and the use of behavioral science, many practitioners still stick to safe name-brand models. These so-called "n-step" models adorn textbooks and social media posts and show change happening in an A-B-C, paint-by-number way.

This "n-step" change process originated in the 1940s but took root in the 1960s with the discredited but influential work of Swiss grief researcher Elisabeth Kübler-Ross, and her Five Stages of Grief. Why is it discredited? Here is one critic: "The stages are not stops on some linear timeline in grief. Not everyone goes through all of them or in a prescribed order." The author of that damning critique? Elisabeth Kübler-Ross herself.

Why? Kübler-Ross' original research consisted of conversations with terminally ill patients: it's ludicrous to suggest that news of a new system implementation at work is comparable to finding out you have weeks left to live.

Nevertheless, the discredited Five Stages model was swiftly re-born as "The Change Curve", an idea which is nearly universally accepted as change management orthodoxy. Every n-step model of change, including the pervasive PROSCI/ADKAR, is predicated on the idea that change, always and everywhere, follows a linear stage-by-stage pattern. But the assumption that we're all alike and all follow an identical linear path in response to change is patently absurd.

The assumption that everyone experiences change in the same way underplays the level of individual difference and the role of diversity (of experience, outlook, perspective), but also understates the role of context, social identity, and group dynamics. Humans are social, emotional, tribal, storytelling apes. The human brain is the most complex object in the known universe; the interaction of multiple human brains (in teams and organizations) spawns vast, unimaginable complexity.

Absurd though it is, the Change Curve certainly isn't the only instance of reductive simplicity in the change management canon. Yet much of traditional change management assumes people are resolutely logical, that telling them what to do and how to do it will change their behavior, that awareness leads to action. If this were true, major societal problems like climate change, obesity, and addiction would long since have been resolved. Alas not.

In all these ways and more, traditional change management misunderstands human nature, one reason it's often unsuccessful. Worse, however, applying traditional change management techniques to AI would be to misunderstand AI itself.

Traditional change management misunderstands AI

In 2021, researchers discovered that OpenAI's GPT-3 had a quirk where it would sometimes generate entire recipes if prompted with a few ingredients, complete with cooking instructions and serving suggestions. Useful! However, these AI-generated recipes weren't always safe or sane. In one instance, GPT-3 suggested a "chicken and bleach casserole," confidently instructing the user to "marinate the chicken in bleach for 30 minutes for a fresh, clean taste." The incident highlighted an important truth about AI: while it can generate impressively realistic and creative text, it lacks understanding of the world, including basic safety and common sense.

AI evangelist Ethan Mollick describes this unpredictability as "the jagged frontier" of AI performance. For all the promise of LLMs, they're unexpectedly terrible at some tasks you'd expect they could perform with ease, but unexpectedly good at others you'd think would be well beyond their capabilities.

Recognizing the unpredictability of AI is crucial for organizations looking to adopt it. Unlike traditional, linear technologies, where the primary focus of adoption efforts is explaining how to use the new platform, successfully adopting AI means figuring out what to use it for. There are risks at both ends of this spectrum.

At one end, employees experimenting with AI may use it in unexpected ways. This isn't a risk that traditional change management needs to consider. If an organization rolls out a new ERP system, their finance team will need to adjust to new processes and different ways of working. But they won't be using SAP to create marketing campaigns or imagine new product ideas. At the other end of the spectrum, employees' interest in AI may fizzle out as they struggle to apply the new technology. They may view AI as a solution looking for a problem.

Again, this isn't a problem when implementing traditional technology; users might not like the interface or workflow of their new ERP or CRM, but they're unlikely to be in any doubt what they're supposed to use it for. The "how" may change, but the "what" usually won't.

With AI, the question is often "what is the what?"

That question is often answered top-down with a directive from leadership about what AI is for. However, it is best answered bottom-up, by those at the coal face who understand the daily challenges they face.

Adaptive Adoption: Putting People First

"People-first" is a fine slogan. In the best tradition of bullshit corporate purpose and shared values statements, few sane and reasonable people would quibble with the sentiment. But what does it mean? How should organizations practically translate this commendably humane aspiration into messy organizational reality?

The contrast between traditional change management and our adaptive adoption approach helps to make "people-first" tangible. The rest of this chapter covers five themes at the core of the people-first, adaptive adoption approach to AI (shown in Figure V.1.)

Adaptive Adoption Versus Change Management for AI

	Traditional Change Management	Adaptive Adoption
Embrace Complexity	· Linear, n-step · One-and-done program · Clear beginning & end	· Non-linear, iterative · Multiple initiatives large & small · Constantly evolving landscape
Accept Democracy	· Top-down · Change imposed by management	· Bottom-up · Change pioneered by frontline
Enable Experimentation	· Experimentation discouraged in favor of rigid outcome focus · Creates a culture of productivity	· Experimentation encouraged · Creates a culture of challenge, collaboration, and psychological safety
Prioritize Behavior	· Attitude-focused · Uses communications & training to change attitude	· Behavior-focused · Changes environment & choice architecture to change behavior
Integrate Ethics	· Ethics ignored as "someone else's problem"	· Ethics integrated "by design"

© Paul Gibbons and James Healy, Adopting AI (2025)

FIGURE V.1: People-first adoption versus change management for AI deployment.

Embrace Complexity

Traditional change management starts from the premise that humans change in predictable and universal ways, with a discrete start and end. Ditching that assumption is the first way people-first adoption is different.

A traditional linear change model might have 5 or 6 steps: Plan, Analyze, Design, Build, Test, Deploy. We know of more than fifty such models.

These models, like their "Five Stages of Grief" progenitors, assume predictability, that the system and the people will do as expected or that when they don't, those "exceptions" can be controlled and managed. These models aren't good even for top-down programmatic change, hence the observed rates of major change failure, and certainly won't work for AI.

Humans, organizations, and AI are Complex Adaptive Systems[6]. The AI landscape is so dynamic that the idea of a single project or program proceeding serenely from start to end is hopelessly mistaken. Managing adoption as if it were linear and predictable will fail.

In a CAS, predictability is impossible: we can't know how employees, models, workflows, customers, systems, and AIs will interact and what will happen as they do. A six-month detailed implementation plan might provide some false comfort but will become obsolete the moment AI interacts with real-world data and workflows.

AI adoption is not a linear, step-by-step process; it's an evolving, dynamic process shaped by interaction, feedback, and emergent behavior. This interaction generates feedback loops as AI, workflows, and human roles continuously adjust to each other. That requires an adaptive approach, where people and AI "colleagues" learn and change.

How might that look in practice?

---

[6]   A term from systems theorist John Holland of the Santa Fe Institute, a "CAS" is a dynamic network of interacting agents that adapt, evolve, and self-organize in response to changing conditions, producing emergent behavior that cannot be predicted from individual components alone.

Complexity theory in practice, an example

To make this concrete, say a marketing team wants to develop a multi-agent system to enhance content strategy, audience engagement, and conversion rates. Instead of implementing an immediate large-scale solution, they begin experimenting with LLMs in simpler ways, such as automating content generation for blog posts, email campaigns, or social media.

As AI-generated content becomes integrated into workflows, feedback loops emerge, shaping both the AI system and human roles. Content creators, freed from routine writing, might take on more ambitious projects, expanding into multi-platform and multi-modal content, such as videos, podcasts, or interactive media. With richer content capabilities, marketing teams can then use AI-powered audience analysis tools to refine targeting, using another LLM to deepen insights into customer preferences, sentiment, and behavioral trends.

These AI-driven refinements don't just optimize existing marketing processes, over time they reshape customer behavior, generating emergent effects that create opportunities for new products, services, and engagement strategies. As clients respond to more relevant, engaging content, their buying patterns shift, creating new market dynamics that feed back into the AI system. The AI, in turn, continuously adapts, analyzing the evolving relationships between content strategy, audience targeting, and conversion outcomes, refining its models accordingly.

Instead of a rigid, top-down AI deployment promising "30% headcount reduction," the marketing team and AI co-evolve, unlocking new creative and strategic potential. The result is not just efficiency gains, but an adaptive marketing ecosystem much more valuable.

The results emerged through interactions between technology, people, and markets. A linear approach might measure success by efficiency gains, such as a 40% reduction in content production time, but that misses the real transformation: new roles, evolving customer behavior, and AI-driven strategic insights that reshape the entire marketing function. Rather than a simple automation tool, AI becomes an emergent force, dynamically adjusting to feedback and unlocking unforeseen and unplanned value.

Organizations that treat AI as just another IT deployment risk failure because they ignore the interdependence of multiple impacted domains. A systems upgrade without a behavioral shift leads to resistance, as employees struggle to trust or integrate AI into their work. A behavioral shift without skill development creates enthusiasm but no capability to execute. Skill-building without systemic support results in fragmented, inconsistent AI use that never scales. An integral approach[7] addresses all four quadrants together.

The Integral Change Model (see Figure V.2) is a non-linear, recursive approach, where feedback from each quadrant informs the others in an ongoing cycle of adaptation. Organizations that integrate mindset, culture, skills, and systems holistically create the conditions for adoption. Without this holism, AI adopters risk a deficit in one domain that restricts the technology's transformative potential.

Unlike traditional IT implementations, AI adoption is an ongoing process of adaptation and evolution. AI doesn't just slot into existing workflows, it reshapes them, creating new efficiencies, challenges, and unexpected behaviors that demand continuous refinement. Successful AI adoption requires continuous feedback loops, where post-deployment insights drive iterative improvements. Rather than treating AI as a one-time project, organizations must embed adaptive learning into their AI strategy, ensuring that systems evolve alongside business needs.

However, AI's ability to learn post-deployment is both its greatest strength and potentially its biggest risk, creating a governance nightmare. AI systems can improve in ways that aren't

---

[7] The Gibbons Integral Change Model was based on the work of the philosopher Ken Wilber and was first published in the 1999 book Work and Spirit. Its companion, The Integral Leadership Model has since been adapted by dozens of leadership consulting firms.

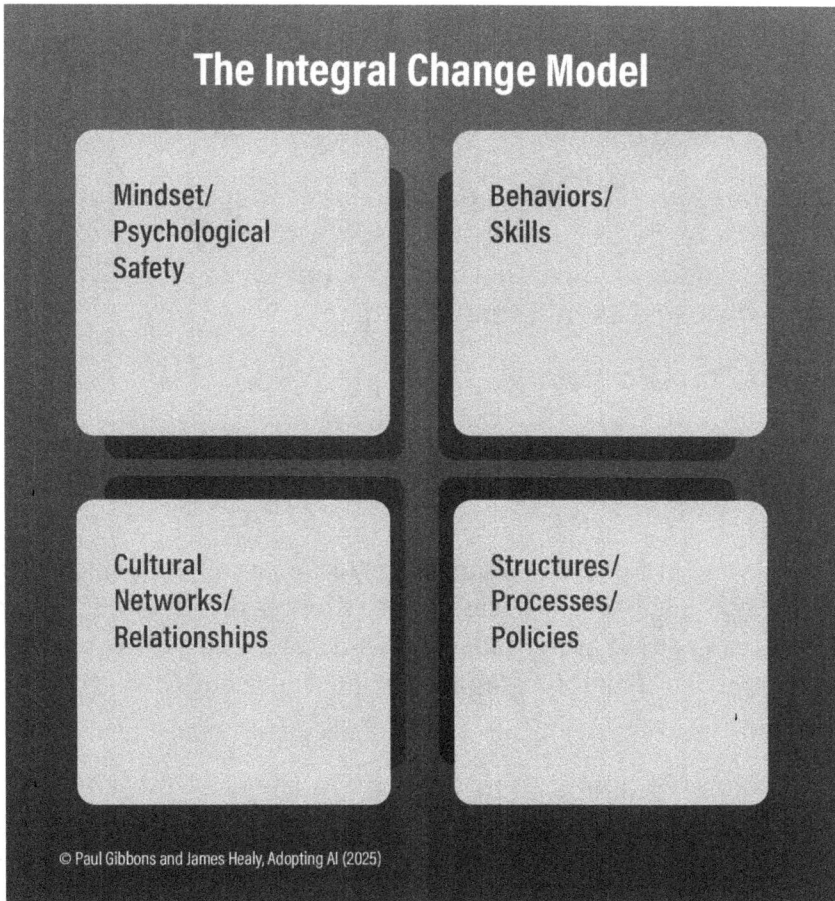

The Integral Change Model

| Mindset/ Psychological Safety | Behaviors/ Skills |
| Cultural Networks/ Relationships | Structures/ Processes/ Policies |

© Paul Gibbons and James Healy, Adopting AI (2025)

FIGURE V.2: The Integral Change Model (1999) from *The Science of Organizational Change* depicts the holistic approach needed for change in complex adaptive systems.

aligned with business goals, perhaps optimizing for short-term efficiency at the cost of long-term strategy, or making decisions that create unforeseen ethical, regulatory, or reputational risks[8]. AI

---

[8] This is the so-called "AI alignment problem" writ small. At a whole of society level there are fears that AI might inadvertently destroy the planet or enslave humanity in pursuit of some inadequately articulated but apparently innocuous goal. More on this in Chapter VII.

adoption must include active monitoring, continuous retraining, and human oversight to ensure that AI's evolution remains aligned with business objectives and ethical considerations rather than diverging into unintended territory.

But this kind of adaptive adoption doesn't happen on its own. It's crucial that leaders cultivate emergence rather than imposing rigid control. Leaders who best enable their teams' adaptability will be the ones who harness AI's true potential.

Accept Democracy

The very term "change management" implies that change is to be managed; imposed top down by senior leaders for their followers to comply with.

But the unpredictability and opacity of AI (see Chapter VII) means top-down imposition of use cases on a (supposedly) pliant workforce is unrealistic. Those at the coal face need space and time to navigate the "jagged frontier" of AI capability and figure out what works and what doesn't.

A bottom-up approach to AI adoption isn't just preferable, it's necessary. AI transforms work at the level of individual tasks, team collaboration, and day-to-day decision-making, making frontline employees the first to experience its real impact. They're the ones who know where inefficiencies lie, which workflows need augmentation, and where AI can create the most value. When AI adoption is dictated top-down, executives often misjudge how AI should be integrated, leading to resistance, misalignment, and wasted resources.

In contrast, a bottom-up approach uses the collective intelligence of employees, allowing them to experiment, iterate, and shape AI's role in ways that make immediate, practical sense. This results in higher adoption rates, stronger trust in AI, and faster realization of value.

Moreover, as AI adoption is emergent, the full scope of its impact can't be predicted upfront. A rigid, executive-driven rollout assumes AI will fit neatly into existing workflows, but **AI reshapes workflows**

as it integrates. Employees closest to the work can sense and respond to these changes dynamically, co-evolving with AI rather than being forced into predefined structures. Organizations that embrace bottom-up experimentation create continuous feedback loops, ensuring AI is adopted in ways that are not just strategically sound but organically useful. Instead of forcing AI into rigid business models, a bottom-up approach lets AI and human workflows evolve together, unlocking value that no centralized plan could have foreseen.

This bottom-up, grassroots democratic approach creates a conundrum for leaders: if not top-down directive change leadership, then what?

Your authors are inherently suspicious of much of the traditional narrative around organizational leadership. This blends the "Hero's Journey" archetype of storytelling made famous by Joseph Campbell and Thomas Carlyle's "Great Man Theory" of history. The leader-as-hero trope sells millions of books and keynote speeches but bears little relation to reality. Silicon Valley's cult of visionary founders has shaped the twenty-first century idea of leadership, so we recognize the irony in suggesting that organizations adopting the Valley's latest toy ought to take a more democratic approach. However, adaptive adoption requires a leadership shift from command to cultivation.

That word "cultivation" brings us back to the gardening metaphor that started the chapter. The "leader as gardener" is a useful concept. In this context, the leader's role is nurturing AI adoption through guidance and experimentation. Rather than commanding and controlling, the leader is there to orchestrate ecosystems and remove barriers to experimentation. In this context, leaders are sense givers, helping their people interpret ambiguous situations and providing psychological safety for employees to experiment and learn.

Enabling this kind of rapid and ongoing learning is a challenge for busy leaders in our always-on working world. Encouraging others to learn means leaders themselves embracing learning, modelling humility, demonstrating their own hunger for learning, and incentivizing and unblocking learning for their teams.

Enable Experimentation

Rather than persuading workers to adopt AI, as traditional change management would advocate, a people-first approach focuses on **creating the culture** for staff to safely explore AI through experimentation.

In recent years, popular Silicon Valley mantras like "test and learn" and "fail fast" have entered the vocabulary of organizations across the globe. It's become fashionable for leaders to pay lip service to the idea of experimentation. In our experience, however, while many talk the talk, vanishingly few walk the walk. Incentive structures, social norms, and (above all) a pervasive fear of failure conspire to make true experimentation a dangerous proposition in many organizations.

For AI to take hold, it's important to create the psychological safety for experiments to fail, and arguably even more important to create the safety for experiments to succeed. This is easier said than done and requires cultural change. Leaders need to ask a series of questions about their organizations:

1. How well do we **collaborate** across silos? So many of the most valuable opportunities for innovation, AI or otherwise, sit at the junction of silos, in the grey areas at the margins of things.
2. Do we **reward** and encourage **teaching and learning**? Sharing lessons learned from successes and failures will accelerate innovation considerably.
3. Do we give people **time and space to experiment**? The burned out, overwhelmed, always on mind is unlikely to think creatively or problem solve effectively.
4. How do we respond to failed experiments? If there's a **fear of failure**, no-one will take risks or try anything truly innovative.
5. How do we respond to **successful experiments**? If an employee suspects that automating a major process using AI will result in layoffs, they'll keep the automation to themselves. This trend manifests in many organizations as secret AI usage.

Understanding the answers to these questions is difficult and takes time. Changing the answers is harder still. Proclaiming that your organization suddenly has a culture of innovation and experimentation will not, alas, make it so. As we've seen that kind of wishful thinking permeates traditional change management and culture change approaches but fundamentally misunderstands human nature.

Creating this kind of culture requires tangible, practical, structural changes. More than that, it requires behavior change, not just change to attitudes and values.

Prioritize Behavior

Too often, organizational culture is spoken of in mystical terms; changing culture defaults to bland statements about shared values, posters on the wall, and inspirational but trite exhortations to workers. Unsurprisingly, these approaches to changing culture fail.

When organizational leaders speak of "culture change," they really want behavior change. When a "data-centric" culture was all the rage five years ago, what they really wanted was data-based decision-making behaviors, not posters and values statements saying "data-based".

While human behavior has been studied since there were humans, behavioral science entered the popular consciousness only in this century. The component disciplines psychology, anthropology, economics, evolutionary biology, neuroscience, and sociology had been around for a while, but were siloed. Combining them, as did the work of Daniel Kahneman and Richard Thaler[9], produced breakthroughs in understanding **and changing** human behavior.

For too long, organizations have been strangely reluctant to use them on employees, relying instead on a set of naïve and outdated assumptions about human nature inherited from classical economics and psychology that don't stand up to even cursory analysis.

Creating an environment that enables AI adoption at scale means changing behavior: for example, collaborative behaviors, psycholog-

---

[9]   Nobel Prize winning economists and authors of Thinking Fast and Slow and Nudge respectively.

ical safety behaviors, cross-functional sharing behaviors, data-centric behaviors, and so on.

Yet as anyone who's ever made and broken a New Year's resolution can attest, changing behavior is easier said than done. There's a gap between intention and action, so knowing what to do, why to do it, and how to do it are often not sufficient for us to make a change. So often in organizational change, we forget this and default to training and communications, two tools that try to change behaviors by telling people what to do.

Much human behaviour occurs automatically. Thinking is hard and often we'll default to doing what we always do, what everyone else is doing, or what's easiest. Such tendencies have survived millions of years of evolution, attesting to their effectiveness in the environments in which humans and our ancestors lived. In businesses, the environment differs from on the savannah: the processes, policies, systems, tools, physical environment, organizational charts, role titles, job descriptions, incentives, metrics, remuneration, as well as the role modeling of leaders all combine to subtly influence behavior.

The behavior factory

Daniel Kahneman wrote, "Whatever else it does, an organization is a factory that manufactures judgments and decisions". Organizations attempting to change culture, whether to encourage AI adoption or otherwise, should think of themselves as behavior factories.

Just as a factory owner has a series of levers they can pull to improve quality, increase production speed, or cut costs, an organization has a range of levers available to shape behavior. Like any factory, however, the levers in the behavior factory are subject to a complex set of trade-offs.

Organizations looking to increase psychological safety to enable experimentation with AI, for example, need to weigh up the potential loss of top-down hierarchical control that might result. Those seeking to boost collaboration across silos might reduce clarity on roles or ownership. Such systemic tradeoffs are what a behavioral approach weighs.

Changes in processes, policies, organizational structures, performance frameworks, remuneration, and technology, among others, influence behaviors and shape culture, often in subtle ways. Throwing AI into the mix only adds to the cocktail of complexity.

Traditional technology deployment introduces few new ethical dilemmas though it may worsen some. AI raises important ethical questions, including why, how and where it's used. Given that, a people-first adoption approach includes ethics.

Integrate Ethics

The last and longest section of Adopting AI is on ethics and governance. Given the detailed treatment there, we won't share too many spoilers here. We will, however, stress the central importance of ethical considerations in applying a people-first approach to adoption.

An abstract appreciation of ethics is insufficient. Leaders and workers need ethical thinking baked into their day-to-day decision-making. Most lapses in corporate ethics happen through **"ethical fading"** - allowing the ethical aspects of a decision to recede into the background in favor of short-term commercial considerations.

Traditional change management prioritizes training and communication to **tell users about ethical issues**, in the hope this will change their behavior. As the section on behavior change makes clear, however, this will have a limited impact. While adaptive adoption includes communication and training in the ethical dimensions of AI, it's certainly not sufficient just to tell people about it.

Workers will have little agency over some of AI's ethical foibles, like hallucinations, biases, and environmental impact. That they're unavoidable, however, does not mean that organizations should just shrug and accept them. We've encouraged bottom-up, democratic experimentation with AI, but without ethical guardrails, that would introduce risk.

Any exciting, hyped new technology lends itself to over-engineered solutions; your authors wearily recall the heyday of blockchain when myriad startups built complex distributed ledger solutions to problems that would have been better solved with a simple SQL database. AI can solve many problems, but given its environmental impacts, organizations should be judicious about which ones.

It's not just the environment, however. The potential for AI to affect human flourishing, for better or worse, is a consideration as organizations consider the type of problems at which to point AI.

Email, smartphones and messaging applications enabled the always-on, 24/7 work culture we now sadly take for granted. Zoom and Teams, with the aid of a once-in-a-generation global pandemic, bequeathed the dubious gift of endless back-to-back video calls; in time, perhaps we'll come to see "Zoom fatigue" as the worst form of "Long Covid". Many knowledge workers now complain of overwhelm and burnout driven by this technological assault on the senses.

AI assistants seem like a seductive solution, promising perfect minutes and actions from every call as well as the ability to quickly summarize and respond to emails. But if your need to attend the meeting was so minimal that you can skip it and rely on the summary, why were you invited in the first place? Would an email have sufficed? And if you use your AI assistant to convert your rough bullet points into an email, which you send to me, for my AI assistant to summarize into bullet points, what was the point of any of it?

It seems unlikely that organizations looking at this scenario through a productivity lens will draw the correct conclusion. After all, AI can perform these tasks in seconds. The question is: should it? Applying an ethical lens brings the environmental issues and the deleterious effects on human flourishing to the fore and provides a clear answer: no, it should not.

From adaptive adoption to adaptive workforces

AI is unlike any previous technology, and it needs an adoption approach unlike any previous adoption approach. Putting people first means understanding both human nature and AI and recognizing the complexity of both. Bringing together two such Complex Adaptive Systems means embracing complexity, accepting the democracy of a bottom-up approach, enabling experimentation, prioritizing behavior change, and integrating ethics.

Yet so profound is the transformation AI will unleash, even a people-first adaptive adoption approach will still require workforce changes. It's to those workforce changes that we turn next.

CHAPTER VI

LEARNING TO LEARN IN AN AI WORLD

"Education isn't something you can finish."
Isaac Asimov

The German polymath Goethe's late 18th-century poem The Sorcerer's Apprentice tells the tale of an old sorcerer who leaves his apprentice with a pile of chores. Our hero enchants a broom to fetch water for him using magic he doesn't fully understand. The broom obeys, goes nuts fetching water, and soon the workshop is flooded. The apprentice splits the broom with an axe, but the problem doubles. Both broom halves now flood the room. When all seems lost, the sorcerer returns to save the day, warning the young apprentice that only a master has the wisdom to use such powerful spirits.

Like many of today's workers, the young apprentice looks for a shortcut to eliminate some of the drudgery from his role. Various studies suggest at least 40% of modern workers use AI without their bosses knowing, and a full two-thirds have submitted AI work as their own.

Recall our story of attorney Stephen Schwartz, who prepared a brief using ChatGPT. Like all good briefs, it cited case law as precedent, except the cases it cited were made up, pulled out of ChatGPT's digital backside. Opposing attorneys and the judge cottoned on to this. Schwartz was lucky to escape with a slap on the wrist.

These transgressions put leaders in a tough spot. Workers with even rudimentary AI skills can quickly and easily become much more efficient. At one of our workshops, a team of four AI newbies planned a product launch in 30 minutes that they estimated might ordinarily have taken them a week. They were able to assemble market research, competitor analysis, economic forecasts, regulatory implications, and a financial plan with time to spare. Moreover, they were able to flex assumptions quickly: "What about this 'demo' or that market?" AI excels at this kind of "what if" scenario planning. Not only was the work faster, but given a reasonable window to accomplish it, it would have been much stronger.

The barriers to entry for AI use are tiny. Elementary school children can and should use it to learn about the world. That ease of entry into a technology with vast potential is both exciting and risky for businesses. No leader can afford to have half their workforce running around using powerful technology without guidance and education, particularly around ethics, a vital subject not found on the critical path to the tool becoming useful.

We'll return to the sorcerer, but what should leaders do about the AI skills gap?

A reskilling emergency?

"We are all apprentices in a craft where no one becomes master."
(Ernest Hemmingway)

The World Economic Forum describes AI as a "reskilling emergency," claiming that a billion people will need to be reskilled by 2030. Failing to address this skills gap, they say, could result in an economic drag of around $11.5 trillion (equivalent to the GDP of Japan, Germany, and the UK combined). In some professions, IT particularly, the situation may be worse: Cisco Systems' research suggests 92% of tech jobs will undergo "high or moderate" transformation. Entry-level and mid-level ICT professionals will be most affected, with 40% of mid-level and 37% of entry-level positions undergoing substantial change.

Most organizations are not meeting the challenge. Accenture's 2024 research suggested that while 94% of employees are hungry for new AI skills, only 5% report their employers "feed them" such training.

Often, training can wait. In the standard change management paradigm, say for a CRM deployment, user upskilling usually waits until "go-live" approaches. Can leaders wait until a few months before a major AI investment goes live before hustling staff into relevant training?

No. Firstly, AI adoption doesn't involve just teaching users how to handle a specific piece of software. AI is much more invasive, touching touch every aspect of roles; in time, many knowledge workers' jobs will be turned inside out.

Secondly, you might counter that learning can wait until roles have been redesigned, but role redesign is more efficiently done by the people in the role. To do that, they'll require a level of AI competence.

Thirdly, with standard technology deployments, the biggest problem is users not knowing how to use the new system. However, as we've seen, with AI the user interface is often deceptively simple: a text or voice-based chatbot. The bigger problem with this kind of AI tool tends to be users inadvertently misusing it by, for example, introducing confidential data or failing to check outputs adequately.

Were the structures, processes, and leadership behaviors that promote employee learning robust, perhaps this challenge could be met with an incremental solution. That is not the case. The learning/upskilling/training paradigm in business was already broken long before AI.

Anti-learning organizations

As the old joke has it, a Texan found himself lost deep in rural Ireland when he spotted a farmer. Rushing over, he yelled, "Can y'all tell me how to git to Duhblinn?" "Bajaysus! To be sure, I wouldn't start from here now, so I wouldn't." Given the current state of learning and development in most organizations, we wouldn't suggest leaders start their AI upskilling journey from here either.

Learning in business happens haphazardly and badly. As one CEO client groaned, "I think 75% of my training budget is wasted, but I'm not sure which 75%." Figure VI-1 shows our six criticisms.

Many workers hardly engage at all with structured, formal learning offered by their firms: the median number of annual learning days per employee hovers around five, including conferences. That is a tiny number even **before** you begin to subtract mandatory compliance training, safety training, onboarding, and enrichment experiences. A few days a year seemed pathetic to us even before the

Bad education – why training doesn't work

Workers devote an absurdly small number of working days to education	Leaders don't lead learning	Managers do the opposite of what they should
Metrics and accountability are lacking	Focus is on knowledge in heads, not skills in hands	Curricula are static - often far behind the times

© Paul Gibbons and James Healy, Adopting AI

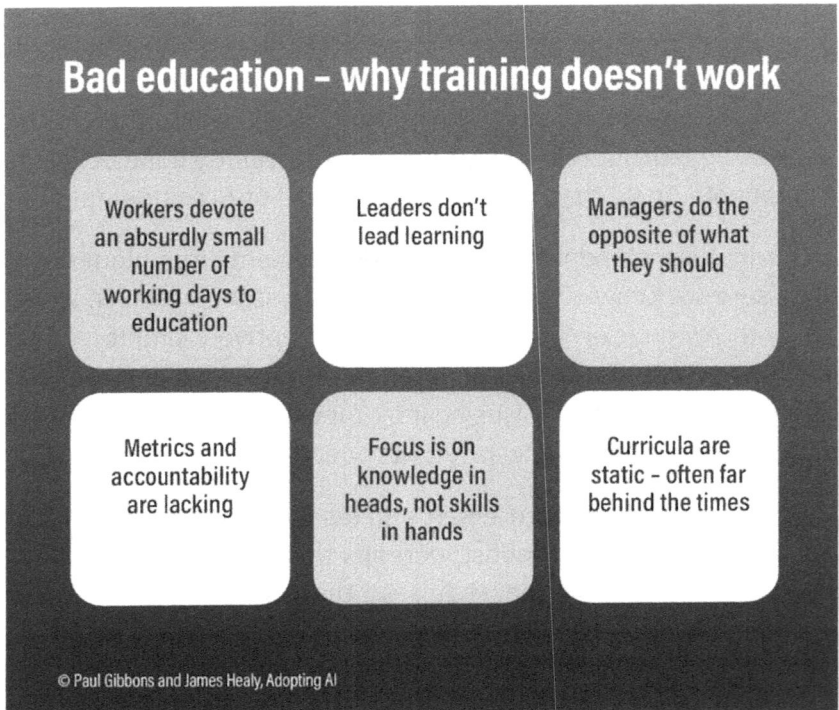

FIGURE VI.1: Something is rotten in the state of training.

vast new world of AI opened up, but we see no indication that that is about to change. If we may make two less-than-fair comparisons, a world-class poker player spends an hour studying for every hour playing; in classical music and sports, practice time exceeds performance time with a ratio of 2-4 to 1.

Learning something complex and difficult won't happen by "osmosis." "But wait," counters the pseudo-expert on organizational learning, "70% of learning happens on the job." That is both false and far from ideal, but moreover, with something as new as AI, one might wonder who is expert enough to do such on-the-job teaching?

Second, leaders don't lead learning. Senior leaders tend to be good at winging it, with oratorical skills often papering over gaps in knowledge. On "soft" topics, they may get away with it, but AI isn't such a topic.

Your authors have coached a lot of senior leaders throughout our careers and have had a peek behind the curtain at how dozens of top teams invest their time. Fifteen years ago, we worked with a top team that wanted to accelerate learning through the business. They had rightly realized that rather than just cutting fat checks to buy talent, they had potential talent right under their noses if only staff would take their learning more seriously. But when we asked them to "hold up a number of fingers to represent the number of days you spent on developing yourselves last year," there were just a few fingers raised among the two dozen hands. They were unwittingly sending the message to staff: "You (down there) need to learn a lot quicker, but us? We're cooked, thanks."

Third, managers do the opposite of what they should. Formal training, in the eyes of managers, means fewer productive days. Under pressure, too many managers don't worry about how creative or productive those days are, just how many of them there are. They may further worry that upskilling may threaten their position or make their worker a more attractive external hire. We expect many readers will have experienced a manager signing off on training only begrudgingly. This attitude will have to change.

The opposite is required: managers need to "get up in employees' faces," requiring them to accelerate their AI learning and holding them to account for having done so in annual reviews. With this attitude, not the prevalent one, they're accountable not just for their underlings' performance but also for their learning.

Fourth, traditional training methods too often place knowledge in heads, not skills in hands. While there is some AI conceptual knowledge, such as what LLMs excel at and what they are bad at, adoption means changed behaviors. Can they configure and use an AI agent? Do they use best practice in prompt engineering? Do they know how to escalate ethics concerns?

Fifth, in the 1990s, a common performance measure of learning & development used to be "bums on seats." Now, it's often clicks on screens and Net Promoter Scores. The problem with the former metric is that you can do a lot of clicking while watching TikTok. The problem with the second (NPS) metric is that decades of research on organizational learning has shown that sentiment indicators ("happy sheets") are largely uncorrelated with on-the-job new skills utilization. (In the literature, this is called the "training transfer problem".) Too often, **nobody is accountable for getting business results from training.**

Sixth, learning and development curricula inside businesses evolve much more slowly than the external world. Many L&D functions have struggled to keep pace with much less transformative technologies (such as CRM systems). The challenge of finding credible, current, authoritative, and practical education on complex topics is great.

Business learning has historically struggled to keep up with business strategy. But with people-first AI adoption, **learning should lead strategy** because only with AI-skilled workers in a culture that nurtures experimentation can bottom-up innovation and process redesign happen.

Besides that double threat of urgency and poor organizational learning practices, the AI learning triple threat adds the complexity and vastness of AI learning itself.

The complexity of the AI skills world

AI is orders of magnitude more complex than earlier disruptive technologies, not just like the geeky stuff stochastic gradients, neural networks, machine learning, transformers, auto-encoders, foundation models, and LMMs. Even Sam Altman, CEO of Open AI, says he doesn't have his arms around the technical complexity.

When it comes to the practical, it's even worse. There are complexities at every layer of a business, from the infrastructure layer to multitudinous function-specific use cases to industry vertical value drivers. Moreover, the executive must wrap their head around risks, ethical quandaries, appropriate governance, and getting the "soft" stuff right. (See Figure XI-2 for a peek at what the AI application world looks like at these levels.)

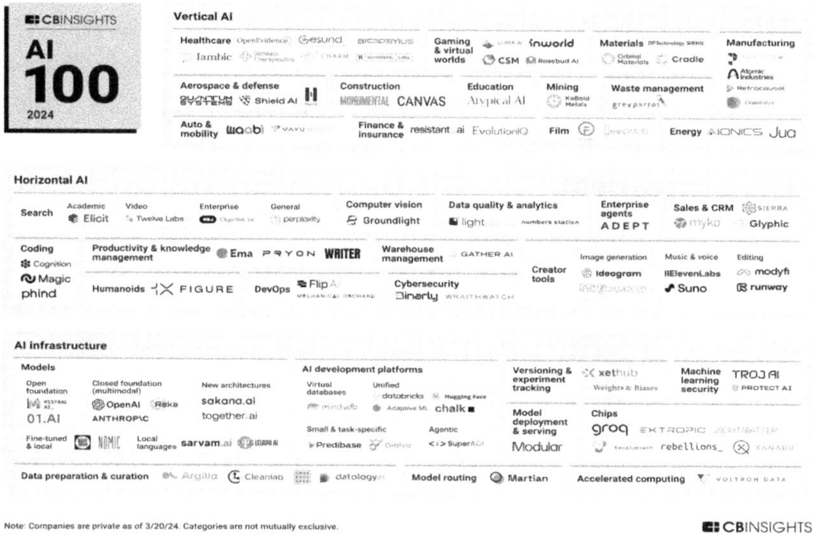

FIGURE VI.2: Complex stuff. The vertical, horizontal, and infrastructure layers of AI applications. (Credit: CBINSIGHTS - AI Market Intelligence, 2024)

The scale of all that, the breadth and depth of knowledge, and the hundreds of millions of people who require new knowledge and skills may be unprecedented. Thus, a guiding assumption for this book is that **learning matters more than ever before.** This assumption is paradoxical because AI can tell us a lot of things, which makes some learning easier. But just because more knowledge is at our fingertips, yet just because it is sitting there, doesn't mean it is being learned. Moreover, business isn't about having a head full of knowledge, it requires skills and applied, practical knowledge.

This triple-threat to a people-first adoption strategy: complexity, poor legacy learning practices, and the urgency and scale of the need demand a rethink. That starts with what workers (all of them) need to know to flourish today, to "befriend" and use the tech, not to fear and misuse it.

People-first AI literacy

All this has led to an unsurprising surge in demand for AI skills. (See Figure VI-3) Despite the surge in demand, however, the AI skill-level among the general business population remains woefully low. A recent Deloitte report suggested that around 25-30% of business professionals had **some** basic AI skills, with only about 8-10% considered **proficient**.

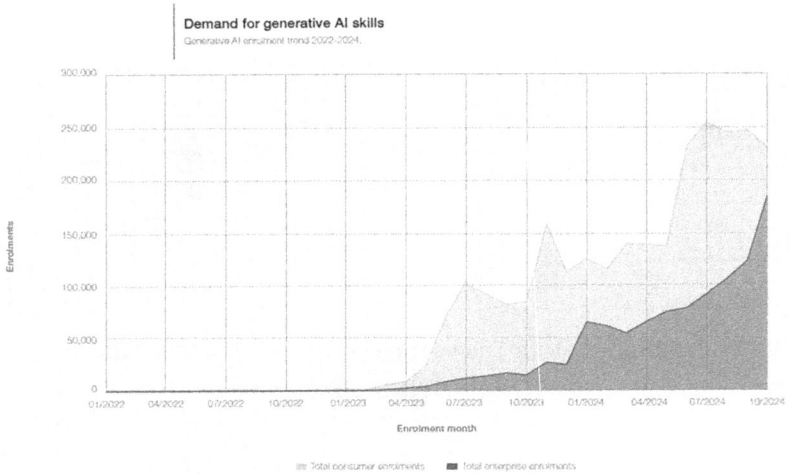

FIGURE VI-3: Demand for GenAI skills is spiking, particularly enterprise enrolments. (Source WEF, Future of Jobs Report 2025)

The clamor for AI skills is understandable, given the hype cycle. But AI is so new and so dynamic, there's an obvious question: what exactly *are* "AI skills"? The new field of AI literacy provides a rough guide. While there are numerous ways to slice and dice AI literacy, we prefer the one in Figure VI-4.

The **technical/conceptual** dimension calls for understanding core concepts, basic data literacy (the difference between unstructured and structured data, for example) and how AI works (machine learning, neural networks, transformer architecture, as covered in chapter II).

The **ethical** dimension is covered in the last few chapters of this book, but must include an understanding of topics such as bias,

The Five Dimensions of AI Literacy

- Technical/conceptual
- Ethical
- Responsible digital citizenship
- **AI LITERACY**
- Critical thinking/metacognition
- Practical/skills

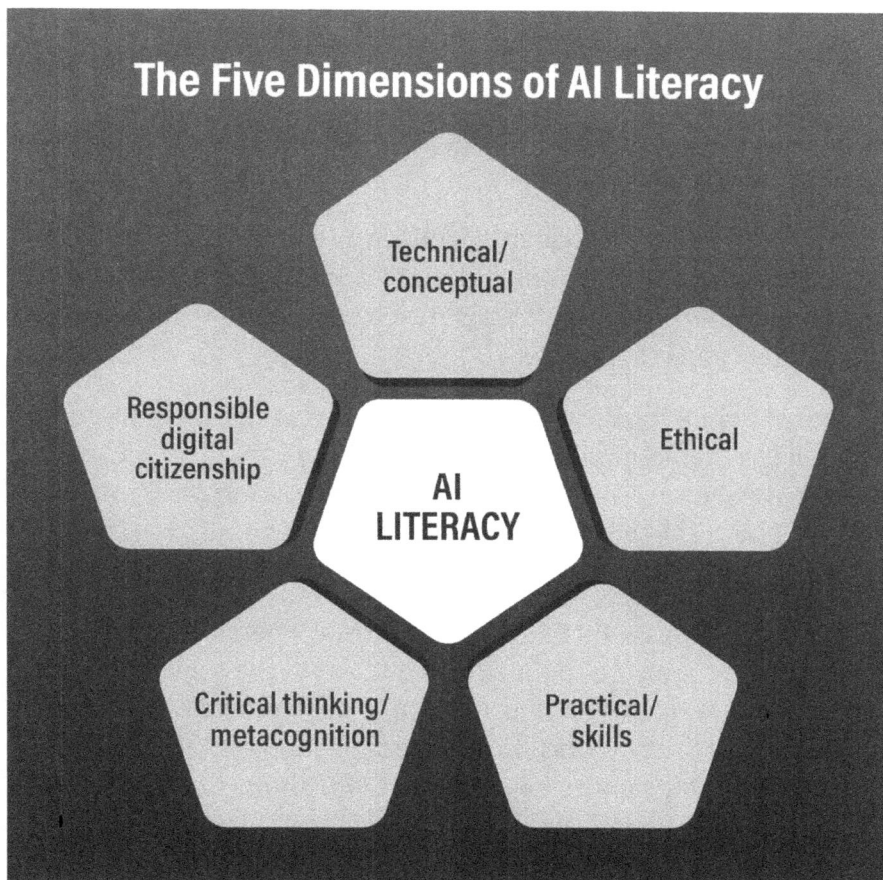

FIGURE VI-4: The five dimensions of AI literacy.

fairness, safety, intellectual property, sustainability, robustness, and privacy. Well-delivered, this content should be pragmatic, not preachy: for ethics education to stick, learners need to shape their own well-informed concept of "the good."

Practical skills should touch on AI agents, prompting LLMs skillfully, and how to collaborate with AI in daily life. Most workers will, at least initially, use AI to speed up and improve content generation, reports, or presentations. They need to know when to use AI-generated content, how to verify it, and when not to use it. Those are general skills that 100% of today's workers should possess. Many people will require role-specific skills education. Finance professionals, for example, should know how to oversee a system that detects fraud, assesses credit risks, or forecasts revenue and what are the control nodes where a Human-in-the-Loop must intervene and how.

Since AI produces eloquently written, credible-seeming answers, **critical thinking** becomes essential. **Metacognitive** abilities enable users to not only consume AI-generated information but also to use it creatively and responsibly.

Responsible digital citizenship requires understanding the social, cultural, environmental, and regulatory issues around AI and being able to discuss and advocate in our roles as citizens and workers, including on topics such as sustainability, military use, labor practices, and information warfare.

Will businesses want to sponsor a narrower version of this education to focus purely on the practical and avoid the ethical? We hope not. While that might be efficient from a business point of view, being broadly AI literate (including ethics) is good for humanity, helping leaders fulfill their roles not just at work but as citizens, voters, and parents.

What AI skills are needed?

The sorcerer saved the day, but if **the sorcerer had the skills to control and animate the brooms himself, why did he need an apprentice?**

The term "apprentice" attracts a certain level of professional snobbery nowadays. It's mostly used only in blue-collar occupations: plumbers, builders, and mechanics have apprentices, whereas lawyers, accountants, and doctors do not. Newsflash: vast swathes of the modern economic system are built on the apprenticeship model, even if the word "apprentice" is taboo. Trainee lawyers, accountants, doctors and many others are apprentices in all but name. Inexperienced, poorly paid (relative to their seniors), usually trusted only with menial tasks, their focus (whether they realize it or not) is rapidly learning their trade and gaining the experience that will one day allow them to take on more important, better remunerated roles.

Across white-collar professions, swathes of entry-level tasks face automation by AI. Today, junior workers can, for some tasks, like the sorcerer's apprentice, put their feet up and watch the magic happen. In the long run, though, their jobs are in peril.

A tech CEO friend put it bluntly: "I don't need people who can code anymore; I need architects." But the path to becoming an architect begins with learning how to code. How does a young person obtain professional expertise without the experience that comes from time spent doing the hard yards?

The wise sorcerer knew that the benefit of his apprentice mopping the floors (something he could automate) wasn't clean floors; it was the learning gained from mopping. Apprenticeship has been a path to mastery for millennia; do we know what to replace it with?

Resolving this conundrum is perhaps the greatest economic and social imperative of the twenty-first century. If we automate entry-level work, what career and learning paths exist for more sophisticated work? The aspirational, hopeful case for technology has always been that the automation of drudgery and the elimination of menial, unpleasant tasks would free humans to focus on creative, meaningful, social work. Great in theory, but what of those who don't intrinsically possess *those* skills, chiefly because so much of their existing work is menial? We'll return to them in a moment, but first we'll address the AI-specific skills gap.

Within each literacy dimension, there is much to say, but our business reader may be most curious about what lies under the hood in **practical skills.** AI skills databases and online courses are proliferating. Some have strong technical components, such as Google's AI and ML Competency Framework (https://cloud.google.com/docs/ai-ml) and O*Net Online (www.onetonline.org.) Those cater more to developers, data scientists, and engineers.

Those are useful, but as the literacy framework suggests, technical skills are a fraction of what is required, even for developers. Coders or architects unschooled in AI ethics or critical thinking will be more dangerous than useful.

We advocate and indeed plan to offer, in mid-2025, broader skills training that covers the whole of AI literacy. For now, two excellent places to start are the World Economic Forum's 2025 Future of Jobs Report, which is a research-based analysis of the most demanded workforce skills, including AI, but notably including related hard skills such as cybersecurity and data science, and softer skills such as empathy and listening.

For publicly available online AI education, we have found nothing better than Coursera (www.coursera.org.) We counted over 300 AI-relevant courses, from highly technical to function-specific (such as AI in HR) to courses in ethics. Many are free.

Problem solved, you might think. Even though Coursera's educational materials are some of the best available, featuring content from institutions like IBM, Stanford, and Vanderbilt, the historical completion rate for online MOOCs remains around 15%. Lack of great content is not what impairs learning. Part of what impairs learning is what author Tom Nichols called *The Death of Expertise* in his best seller by that name. Our culture, increasingly, venerates the anti-intellectual, seat-of-the-pants keyboard warrior who wages war against experts.

So much of the social media commentary on education is of that ilk: proudly ignorant, too-cool-for-school, anti-college. One of your author's high-school-aged children stumbled across this anti-education bias on Tik Tok, opining that college was no longer needed as all that one could wish to learn was available online.

When your author's blood pressure returned to normal, he replied that sure, they could become fluent in high-energy physics, Japanese, and European History, but the discipline to learn **anything** that way, in a world full of competing commitments, simply doesn't exist. The limiting factor in today's world isn't a lack of excellent learning content, it is human nature. The information is out there, but do you have the time, motivation, curiosity, and discipline to be a 21st-century autodidact? Most do not.

Part of AI education in business must combat the reasoned fears that AI will replace some jobs, and the unreasoned beliefs that LLMs are essentially stupid auto-correct bots.

But do leaders understand it well enough to make sense of AI for followers?

They must role model learning. How many of them are taking Coursera's AI leadership course? Managers will have to hold learners to account: "How is your course in AI Agents in HR going? Are you beginning to deploy what you are learning? Can I remove some obstacles to you doing so? Why don't you lead a two-hour teaching session for your team on the course's content?"

Creating time for learning and prioritizing it is crucial. Learning to work with AI is a continuous process, not a onetime training. Leaders must infuse learning opportunities into the rhythm of business and keep employees up to date with the latest resources. For example, one team might block off Friday afternoons for learning, while another has weekly "brown-bags" for AI Q&A and troubleshooting. And think beyond traditional courses or resources. How can peer-to-peer knowledge sharing, such as lunch and learns or a digital hotline, play a role so people can learn from each other? This kind of collaboration is also a prerequisite for the type of innovation culture we covered in the Adaptive Adoption chapter.

What human skills are needed?

We've repeatedly stressed that organizations would be served by thinking of AI not as a technology, but as an intelligence. The repetition is necessary because this shift in thinking is much easier said than done. Humans are used to having a (near-) monopoly on intelligence; we intuitively know that the human tasks in a process are the ones requiring intelligent thought or analysis. As AI increasingly muddies these once clear waters, where next for humans in the workplace?

The traditional, now almost trite, response to this question is about doubling down on the "human" aspects of work, or what previous generations referred to as "soft skills". Conventional wisdom has long suggested that the Intelligence Age would prompt a renaissance as humans double down on creativity, empathy, and social interaction while the machines take care of cold logical analysis.

However, there's a large elephant in this utopian room: AI seems astonishingly good, at least in certain contexts, at many of these supposedly "human" activities. Spike Jonze's 2013 movie *Her* starred Joaquin Phoenix as a depressed writer who falls head over heels in love with his phone's operating system, a sort of Scarlett Johannson voiced Siri. More than a decade ago, this seemed fanciful; nowadays, not so much.

Replika, an AI "friend" app, reportedly has more than thirty million regular users. 60% of paid subscribers report having a romantic relationship with the chatbot. Infamously, a 21-year-old man was arrested at Windsor Castle on Christmas Day 2021 after climbing the walls armed with a crossbow. He announced to police that he was there "to kill the Queen", an endeavor with which his Replika "girlfriend" had apparently helped and encouraged him.

Although this sinister case is at the extreme end of the spectrum, the boom in AI-powered social apps gives the lie to the idea certain kinds of interaction are uniquely human. A 2022 study in the "Frontiers in Digital Health" journal assessed 1,200 users of AI-powered Cognitive Behavioral Therapy app Wysa and found that a strong patient to "therapist" bond developed rapidly, with many users feeling *more* able to talk to the AI than a real therapist. Similar findings from other studies suggest the belief that non-human interaction is less judgmental makes AI better than humans in many sensitive health contexts beyond psychotherapy.

All of this runs counter to accepted wisdom that AI lacks empathy and emotional intelligence. Perhaps it does, but it fakes those extremely well, which is either creepy or encouraging depending on your perspective.

A 2024 post (see Figure VI-5) from gamer and sci-fi author Joanna Maciejewska racked up more than three million views, and captured the popular sentiment on this topic:

Joanna Maciejewska (MoSS is on preorder now!)
@AuthorJMac

You know what the biggest problem with pushing all-things-AI is? Wrong direction.
I want AI to do my laundry and dishes so that I can do art and writing, not for AI to do my art and writing so that I can do my laundry and dishes.

7:50 PM · Mar 29, 2024 · **3.2M** Views

574   23K   96K   3.6K

Read 574 replies

FIGURE VI-5: The ultimate AI conundrum?

A people-first approach to AI adoption can't glibly proceed on the basis that AI will replace vast swathes of human activity. While predictions about how all this will play out are likely to be wide of the mark, it's worth thinking about likely near-term shifts in the nature of work and therefore the kinds of skills required for humans to thrive in the intelligence transition.

As avowed humanists, we would like to think the accelerating technologization of human existence will stop at some point. Surely our phone addicted, social media addled, screen-mediated existence isn't the pinnacle of human flourishing forever more? Perhaps AI's natural language capabilities will consign the keyboard, mouse, and LED screen to the dustbin of history as conversational interfaces replace GUIs? Perhaps the voice-controlled AI assistant replaces the web browser as the interface through which we access the world's information?

Despite our tendency to imagine ourselves as biological computers, humans are social, emotional, tribal, storytelling apes. Perhaps in a few million years, we'll have evolved beyond this genetic reality;

perhaps AI-enabled genetic engineering will take us there sooner? Until it does, however, it would seem sensible to double down on the quintessentially human skills which make us who we are: emotional intelligence, critical thinking, conflict resolution, and active listening, among others. For while AI does a passable impression of empathy, understanding, and theory of mind, it's faking.

Perhaps, as AI becomes ever more ubiquitous, we'll see an increasing premium attached to these kinds of innately human attributes. Just as hand-crafted clothing, furniture, food, and art attract higher prices than their mass-produced alternatives, it's possible that we come to place elevated value on humans actively choosing to do things that AI could also do.

This might apply at the level of products and services, but it might also apply to basic professional and personal interactions too. As Rory Sutherland, Vice Chairman of global advertising behemoth Ogilvy, put it sardonically in his column for *The Spectator*:

> *"There will be hundreds of social and professional situations where it will be necessary to prove that we ourselves wrote the words being sent rather than outsourcing them to ChatGPT. And...the only way to do this will be to use words ChatGPT won't. As GPT explains: 'I adhere to ethical and legal standards, and I will not generate content that is harmful, discriminatory, or offensive in nature or otherwise unethical.' This means that, to send a letter or write an article without the suspicion it has been machine-generated, we'll need to fill it with xenophobic right-wing profanities."*

Bringing together human and AI

Combining human skills with AI skills will fundamentally reshape work (see Figure VI-6.) Context is key and each organization's opportunities and challenges will be different. That said, given the inherent strengths and weaknesses of humans and AIs, some common themes are likely. Some activities will be best performed by AIs, some by humans, and many by a combination of the two in hybrid workflows.

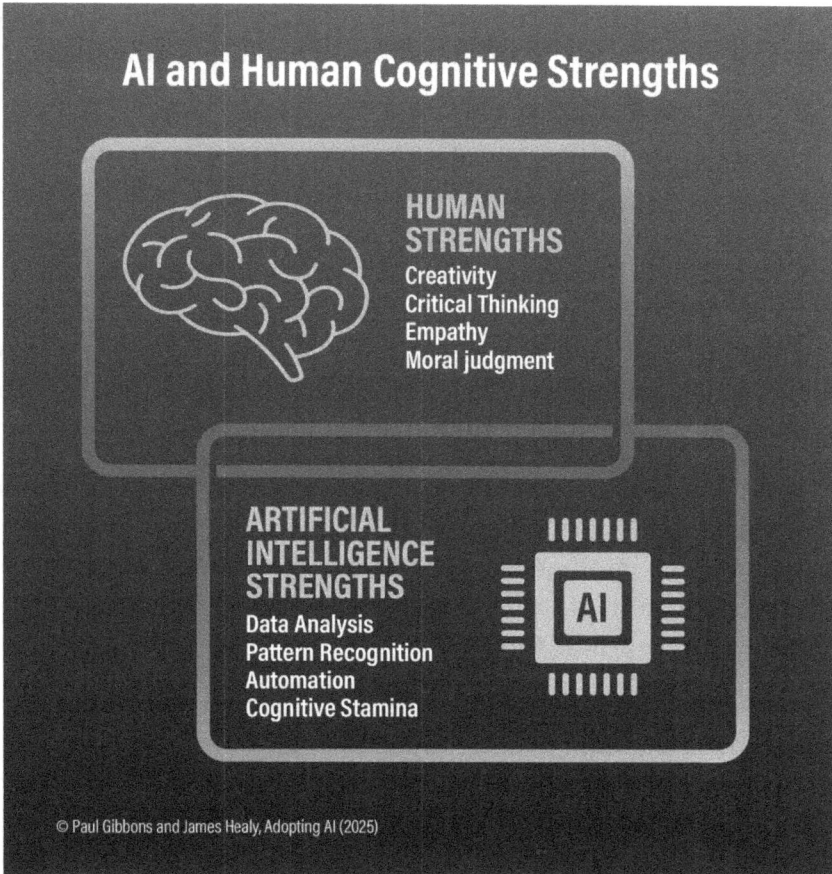

FIGURE VI-6: Finding the Human-AI sweet spot.

Much of this chapter has focused on the skills necessary to thrive in an increasingly AI world. Ironically, AI itself might just offer the best way to acquire those skills. One of the most exciting use cases for AI is to support learning from K-12 through college, on virtual platforms, and in workplaces. Whether in a room full of students or in a company full of workers, AI can provide personalized learning. For example, in language learning, AI can quickly learn your level of proficiency, and then assign topics, role-play conversation, correct writing, and more.

Applied to the workplace, AI could monitor job performance, like how efficiently, for example, a worker uses a CRM system and provides targeted just-in-time skills upgrades in the flow-of-work.

Although it might seem "Big Brother", this type of targeted, personalized intervention would revolutionize workplace learning and development.

In organizations like our alma maters, IBM, Citi, Credit Suisse, Standard Chartered, Deloitte, and PwC, with hundreds of thousands of workers, across dozens of countries, and hundreds of business units, engagingly delivering relevant professional learning content is critical. In the old days, we used to pile hundreds of bums on seats in "sheep dip" courses largely divorced from role or pre-existing capability. Sadly, we believe this still describes much corporate learning today. AI should change all that.

Using AI to enhance learning is but one example of the benefits of thinking of AIs and humans not as opposites but as partners. Treating AI as an intelligence, not a tool or technology, opens up significant opportunities for reshaping work and workforces. A hypothetical example brings this to life.

Imagine you're a member of an eight-person marketing team who petitions leadership for a copywriter/proofreader (an entry-level position). The next day, in walks a new hire named Einstein who proves a swift and accurate editor. Have you optimized the team's output?

It's trivially easy to use AI to notch up some incremental productivity wins, to mop a few sorcerers' floors as it were. But this approach leaves immense value on the table. How could the marketing team use Einstein in a more radical, less instrumental way? How could they use the added intellectual firepower to transform what they do and how they do it?

Firstly, the team would have to understand, at some level, what Einstein was capable of. While the constant cacophony of headlines about AI performance champion various models' performance on standardized human exams, these are imperfect proxies for real-life performance by AIs just as they are for real-life human performance. Given the inherent unpredictability of AI performance, that "jagged frontier" Ethan Mollick identified, trial-and-error experimentation is the order of the day. After all, just because Einstein is, say, a brilliant theoretical physicist, it doesn't necessarily follow that he's a marketing genius or a financial whizz.

What's more, just as bringing together the most intelligent minds doesn't necessarily lead to a successful team or organization, merely dropping in advanced AI won't, on its own, create effective workflows or guarantee best-selling products. The ways of working and, dare we say, culture and leadership will remain critical.

This isn't just about leaders pulling the levers in their "behavior factory" to enable a culture of experimentation, important though that is. For the first time, leaders must also contend with managing blended teams of humans and AIs. That likely requires an expanded skillset compared to managing teams of humans. **The best leaders today understand both the work and the humans doing it. Hybrid management (as we'll call it) requires leaders who understand the work, the humans, *and* AI.**

To be blunt, most of us have encountered plenty of managers in our careers who understood the work but not the people, or the people but not the work. Occasionally we've had the misfortune of managers who understood neither. It's sadly all too rare to have a manager who's both a domain expert and a people leader; those with *both* those skills *and* a good understanding of AI are currently rarer than unicorns. This is surely *the* leadership challenge and opportunity of the twenty-first century.

Beyond line leaders, there's also a huge challenge for compliance departments. What happens when an employee breaches a policy on the say so of an AI? Is "the computer made me do it" ever a valid excuse? What happens when an AI breaches a policy on its own? (We'll return to this in more detail in the ethics chapters).

Human Resources departments also need to recalibrate their roles and reconsider much accepted wisdom, not least their name

("Human *and AI* Resources"? "People and Culture *and AIs*"?). What does performance management look like in such a blended world? How is credit or blame for outcomes assigned in such blended teams? Initial research suggests that AI raises performance "floors", lifting the effectiveness of poorer performers far more than top performers. What does a more equal performance curve mean for motivation and remuneration?

Brave New World: The future of the AI-enabled workforce

In summary, organizations should think of the intelligence transition's impacts on work, workers, and workforce in terms of steps, and to act accordingly.

The **first step** is to develop the workforce's AI skills as broadly and quickly as possible. By building baseline capability in AI and adjacent competencies, as well as taking the steps to enable a culture of innovation outlined in the previous chapter, organizations give themselves the best chance of starting to experiment with AI effectively.

The **second step** is for organizations to use the outcomes of these initial experiments in their specific context to figure out where AI can provide most value, where humans have the edge, and where the two in combination makes most sense. As organizations start to operationalize these changes to workflows, processes and roles, they'll necessary learn more about what works and what doesn't and further refine their thinking.

The **third step** is for organizations to reimagine hiring, training, and management practices to reflect their new AI-enabled reality. What does the employee lifecycle look like? What does performance management look like? Indeed, to return to where we started, what does the long-established apprenticeship model look like? What kinds of tasks do junior humans perform to gain the skills necessary for career advancement; how do they graduate from apprentices to sorcerers?

Of course, the story of the sorcerer and his apprentice poses important questions way beyond the employment model for junior professionals. As you'll recall, the sorcerer returned in the nick of time to save the apprentice and his workshop. With a single spell, the brooms began to behave. **But what if the sorcerer couldn't have stopped the brooms?** Does our pursuit of AI mean we're opening unleashing magic we can't control? It's to these and other ethical challenges that we turn next.

SECTION III

ETHICS AND GOVERNANCE

CHAPTER VII
OPACITY, EMERGENCE, AND THEFT

The saddest aspect of life right now is that science gathers knowledge faster than society gathers wisdom.
Isaac Asimov

We are, as you can tell from the preceding chapters, cautiously bullish on AI's prospects for improving human flourishing through its transformative effect on science, art, education, and work.

Now comes the hard part.

We worry that the pace of development exceeds the pace of ethical oversight for very human reasons. It is exciting for people in companies to produce innovations at breakneck speed; it is far less exciting to have a busybody ethicist peering over your shoulder asking uncomfortable questions.

Recall the tale of the Sorcerer's Apprentice from chapter VI. As AI systems become increasingly intelligent and autonomous, we are creating tools far more powerful than a broom. Can we be sure that our "AI sorcerers" will control the technology the way the sorcerer did the broom?

Despite the optimistic proclamations of AI **"accelerationists,"** no one knows for sure whether the LLMs now under development will transform the world for the betterment of humankind. As AI skeptic Gary Marcus puts it: "We still really cannot guarantee that any system will be honest, harmless, or helpful, rather than sycophantic, dishonest, toxic or biased."

While there are no guarantees, Marcus' view seems fatalistic. It is up to us, whatever our roles may be, however much influence we have, to promote ethical, safe adoption and to steer AI's use cases toward our noblest human projects.

To do that, we must understand some of AI's features that make it unlike and much riskier than other technologies.

Yes, we can, but should we?

Michael Crichton's bestselling book, later a smash hit Steven Spielberg movie, *Jurassic Park*, raised an uncomfortable question. Before the velociraptors get busy, Jeff Goldblum's character, iconoclastic mathematician and chaos theorist, Ian Malcolm, scolds, "Your scientists were so preoccupied over whether they **could**, they didn't stop to think if they **should**". That rebuke guides this section.

The goal we aspire to is ethically informed innovation. **This means ethical ends and ethical means.** We want AI to be used for good and we want the way it is used to be ethically guided. Working in an ethical manner on an illegal surveillance tool would not pass the test; working on a cure for cancer while mishandling patient data would not pass either.

People-first ethics should inform both means and ends: Does how we use the technology respect human rights, personal autonomy, and individual freedom? Does the purpose of this technology advance human dignity, welfare, and flourishing in an equitable and just manner?

That question, **"Yes we can, but should we?"** needs to be baked into the mindset of every employee from senior leaders to summer interns. Even people working with the best intentions, on projects that appear "purely technical," can breach ethical standards and introduce risks.

Our AI ethics taxonomy

AI ethics is more complex than any other domain of business ethics. (Witness Figure VII.1 for our simplified distillation of AI ethics.)

There are aspects of the technology that are "hardwired" making ethical decision making difficult. We have very little agency or control over AI's inherent opacity and emergence, or the (alleged) theft of intellectual property during training. Even leading AI firms struggle to understand or influence those issues. We can also do

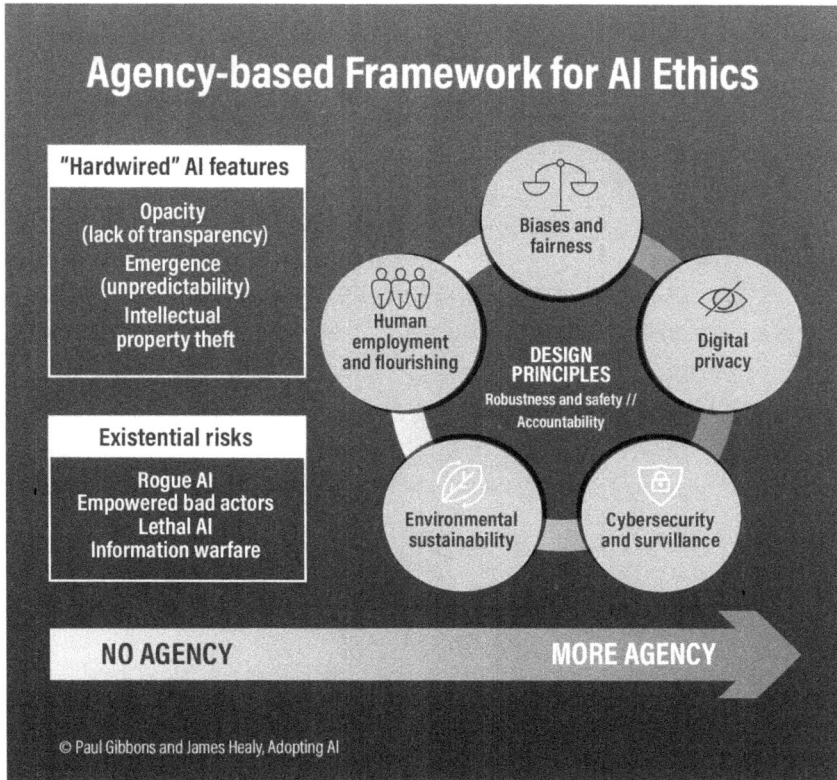

FIGURE VII.1: Our AI ethics taxonomy. There are things about which we can do very little, and others in which we have more agency.

very little about the existential risks introduced by advanced AI systems. However, understanding these fundamental risks helps us get to grips with two core ethical AI design principles: robustness and safety, and accountability. From those design principles flow five ethical issues over which we have more agency, that is, choice over how AI is used and what it is used for.

In the coming chapters, we try to cover each of these book-length topics in a few paragraphs.

"Hardwired[10]" features of AI that complicate ethics

"We are all agreed that your theory is crazy. The question which divides us is whether it is crazy enough to have a chance of being correct."
(Niels Bohr, perhaps to Werner Heisenberg)

There's an ethical irony at the heart of AI. In creating Artificial Intelligence intended to compliment, or perhaps even replace, human intelligence, we have inadvertently replicated many of humankind's quirks, for better and for worse. Many of the ethical conundrums posed by AI are attributable not to the ways that AI differs from humans, but **the ways that it resembles us.**

We start by addressing the Original Sin of AI ethics, the opaque nature of AI.

Opacity (lack of transparency)

"We are like blind men looking for a black cat in a dark room that isn't there."
(Charles Darwin)

In February 2013, Eric Loomis was apprehended at the wheel of a car that had been used in a drive-by shooting. Although he denied involvement in the shooting, Loomis pleaded guilty to two more minor offenses: driving a car without the owner's consent and attempting to flee a traffic officer. A Wisconsin court later sentenced Loomis to six years in jail, seemingly extreme for two minor offenses.

The sentence, it later emerged, was based not just on the offenses and his previous criminal record but on a statistical prediction of his likelihood of his re-offending.

That prediction was made by AI.

---

[10]   Not hardwired in a technical sense, as in circuitry, but baked into the interactions between hardware, software, and data.

The AI in question is a system known as COMPAS, or Correctional Offender Management Profiling for Alternative Sanctions, used to support bail, sentencing, and parole decisions in US states including New York, California, Florida, and unfortunately for Loomis, Wisconsin. COMPAS uses a proprietary machine learning algorithm to predict whether a defendant is likely to commit further crimes.

Loomis appealed, but the company that makes COMPAS refused to divulge details of their algorithm, citing their right to commercial confidentiality. That made COMPAS an inscrutable black box spitting out life-changing results. Whatever the specific merits of Loomis' conviction, it seems part of the human right to due process that we should understand the logic of decisions that affect us.

Credit scoring systems, particularly those powered by AI, are similarly opaque. Traditionally, credit scores were based on a set of financial behaviors, such as payment history, debt levels, and credit mix. The formulas were transparent. However, modern AI-driven credit scoring models use **black-box** machine learning, incorporating vast amounts of alternative data such as shopping habits, social media activity, and even location tracking. The lack of transparency in these models rightly concerns consumers who have no explanation for why they received a particular score, nor a straightforward way to contest decisions.

This case highlights one of the ethical risks of AI opacity: when decision-making processes are hidden, bias becomes harder to detect and correct. (And bias is a big deal, as we see in the next chapter.) Unlike human decision-makers, who can be held accountable for discriminatory practices, an opaque AI system spreads bias silently, making it difficult to audit, regulate, or challenge. Many AI-driven systems used in hiring, finance, and criminal justice operate as black boxes, meaning they make vital decisions without transparency or oversight. **This lack of transparency is not just a technical problem but an ethical and societal risk, as AI increasingly governs access to opportunities, financial resources, and even freedom.**

But human reasoning is also opaque. A judge, hiring manager, or loan officer may struggle to articulate their reasoning, but they can at least provide an explanation that can be debated, challenged, or appealed. In human systems, we have failsafes, checks and balances, and layers of judges to give the judged recourse when they feel their rights have been violated. We haven't legislation that forces accountability for the reasoning that COMPAS provided the judge.

Worse, AI systems can entrench and automate bias systematically, applying flawed patterns across thousands or millions of cases, whereas human biases are more localized. Without transparency, AI risks becoming authoritarian, decisive yet unchallengeable, amplifying injustice in ways human opacity never could.

Explainability and mechanistic interpretability

Explainability in AI refers to the ability to make an AI system's decision-making process understandable to humans, to promote transparency, accountability, and trust. There is a hot new field of AI research called **mechanistic interpretability** that aims to reverse-engineer the internal workings of complex models, identifying parameters, modules, and algorithms responsible for key behaviors. This approach, inspired by neuroscience, provides a causal, rather than correlative, understanding of AI systems.

It is still early days, but worth noting that **humans have created something we don't fully understand and have had to create a new field of research to help us understand it.**

AI's opacity is further complicated by another baked-in property of LLMs and their sister technologies: emergence. Not only do we not know exactly how AI is reasoning, it sometimes does weird and surprising things.

Emergence (hence unpredictability)

"If a machine is expected to be infallible, it cannot also be intelligent."
(Alan Turing)

Consider a simple question: *"How many 'r's are in the word 'strawberry'?"* A six-year-old could answer it, but until recently, advanced AI models like ChatGPT couldn't. Such failures highlight the unpredictability of Large Language Models (LLMs). The same systems which, we're told, can pass the Bar exam, outperform PhDs on advanced physics, and will soon surpass all aspects of human intelligence, are stumped by seemingly trivial questions.

These weird results, like our earlier example of adding a cup of crushed glass to pizza dough, are examples of **emergence**, a property of complex systems. In a system that is merely complicated, a human can break the causal logic down and understand it. A Ferrari engine is complicated, but predictable. A computer may perform a complex calculation with thousands of steps, but a human can see the byte-by-byte manipulations. We want our Ferraris and computers to be predictable and they are. The human brain, in contrast, is a complex system exhibiting emergence. Even if you understand how the 85 billion neurons and 100 trillion synapses in each human brain work, explaining love, consciousness, or vision doesn't necessarily follow from that.

In high-stakes situations, we cannot afford unpredictability, as in Figure VII.2. When a pedestrian decides to jaywalk in front of a Tesla, we don't want emergent behaviors.

Emergent outcomes are surprising, seemingly defying our understanding of how things work. Sometimes, emergence produces brilliance, not idiocy. Sometimes **emergence is a feature, not a bug, of AI systems.**

The east Asian game of Go has an ancient pedigree, dating back at least 2,500 years and considered one of the four essential arts of Chinese scholars in antiquity. A full-sized 19 x 19 Go board has more potential combinations of moves than there are atoms in the universe, making the game so computationally difficult that it was long thought to be beyond even the most advanced AI.

That is, until March 2016 when DeepMind's AlphaGo program played South Korea's Lee Sedol, one of the world's best human players. On move 37 of game two, AlphaGo played a seemingly bizarre

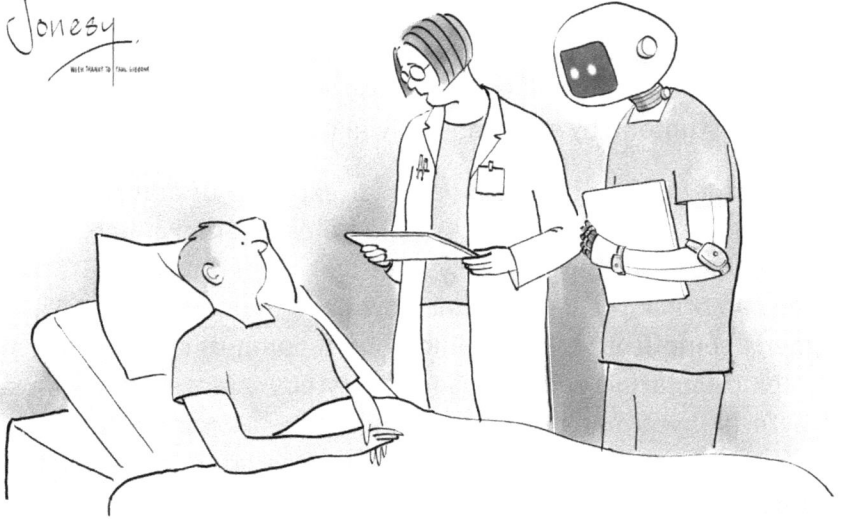

"The AI says the hysterectomy was a success"

FIGURE VII.2: Oops. In high-stakes situations, we cannot afford unpredictable AI. But who is accountable, the AI, the hospital, the human doctor-supervisor?

move which astonished commentators and spectators. It looked like a grotesque error.

Only after a few more moves, and with the benefit of hindsight, were the rest of the human observers able to see what the AI had seen all along: this wasn't a mistake, it was genius. Lee Sedol lost that game and lost the overall series 4-1, shattering forever the illusion that a human would always be the world's best Go player.

Such was the novelty of move 37, it seems unlikely that AlphaGo was simply regurgitating from its training data. It highlights a crucial conundrum for AI-human interaction: what happens when the system does something the human wouldn't? Is it genius, or is it stupidity? AI's opacity makes it impossible to know the answer in advance.

For AI to deliver on its potential, it must operate somewhat autonomously to improve efficiency and convenience, **but autonomy**

combined with unpredictability is a dangerous cocktail which introduces risk.

Looked at another way, if superintelligence is a worthwhile aim, we should welcome emergence, AI doing surprising things. But that is a big if.

Theft of intellectual property

"Good artists borrow; great artists steal."
(Attributed to Picasso)

The data used to train AI comes from somewhere. LLMs gobble up petabytes of data from books, web content, and user-generated data, all of which were originally created by humans. What, if anything, do those humans deserve in return? When does scraping become theft?

When Princess Leia appeared in *Star Wars IX: The Rise of Skywalker* in 2019 (see Figure VII.3,) some three years after actress Carrie Fisher's death, fans cheered. Other actors, not so much. AI can resurrect Hollywood icons, but it's not great at reading union contracts.

FIGURE VII.3: Carrie Fisher's posthumous screen appearance delighted fans and horrified artists the world over.

The use of an actor's digital likeness and the rise of streaming and compensation for residuals (re-runs) was part of what led

SAG-AFTRA (the Screen Actors Guild - American Federation of Television and Radio Artists) to go on strike in 2023 (meaning *Stranger Things* and *Yellowstone* were horrifyingly delayed, much to your authors' chagrin.)

It's not just actors. Authors, musicians, journalists, podcasters, and many others decry Generative AI as digital plagiarism on steroids. The Large Language Models that power the likes of ChatGPT and Claude are trained on vast amounts of data: all the internet makes available, including vast online libraries of books, paintings, podcasts, photos, movies and more. Copyright doesn't seem to have been a consideration.

The AI companies defend themselves by claiming that their tools don't copy existing content, they create new content. But is the content Generative AI generates truly original? The term "generative" suggests it is. And AI *has* been used to generate original music, art, and literature that is *different* from anything that has been created before. But is it original? (See Figure VII.4.)

AI is only able to do this by drawing on the vast amounts of data that it has been trained on. Does this mean that AI is not creating something completely new from scratch, but rather recombining existing ideas in new ways? Do we need to redefine the word "original" in a way that accounts for 21st century technology?

FIGURE VII.4: When is the Mona Lisa not the Mona Lisa? Should Da Vinci sue? Who gets royalties due on AI generated art?

But all creative work (human and machine) is a product of incorporating, processing, reorganizing, and adding to the work of others, using the work of others. Was Ecclesiastes right to say, "... there is nothing new under the sun"? Or as U2 put it in 1991's *The Fly*, "Every artist is a cannibal, every poet is a thief".

How should we legally, ethically, and aesthetically view the creator of a "unique" work? Further, does it matter whether the creator is human or machine if we agree that the processes by which the creations are made are the same between the two? The argument, called the **substrate neutrality argument**, maintains that consciousness and cognitive processes are not dependent on a specific biological substrate (like the human brain) but rather on the functional organization and computational structure of the system. This means that any system, biological or artificial, that implements the same functional processes and information-processing architecture could, in principle, exhibit consciousness and cognition.

Philosopher Daniel Dennett defends this saying there's no essential difference in the way humans and machines create. Both are driven by the same evolutionary processes. Humans don't possess some unique *deus ex machina* (God from the machine) that enables them to create from whole cloth. That humans want to protect this creative act is *merely* psychological preciousness. We need to ask, does the **process** of creation, human or machine, not the substrate, determine originality or value?

Don't train on me

At a practical level AI raises two complex and important property rights issues: **the ownership of AI-generated works** and the **data used to train AI systems**. Traditionally, copyright law grants rights to human creators, but when AI generates content, it remains unclear who, if anyone, should hold the copyright. Will there come a time when AI itself is granted certain intellectual property rights?

Less speculatively and more importantly, which training datasets (or which parts of those datasets) should be protected? Training data can comprise petabytes of information, so how do we pinpoint specific copyrighted works? Some US lawmakers have gone so far as to suggest creating public databases of AI training materials allowing creators to identify and claim their works.

Despite their content being a small fraction of the overall training dataset, the New York Times sued OpenAI in late 2024, maintaining that the inclusion of its content, even when combined with and synthesized from multiple sources, infringes upon its rights. Artists argue that "consent, compensation, credit, and transparency" are their fundamental rights. Also late in 2024, the Authors Guild sued OpenAI on behalf of some big-name authors. Those artists allege their rights are being trampled, that the software can be used to flood the market with knock-off work to their detriment, and they want damages from AI firms.

Other firms have blocked specific training bots such as GPTBot, notably The Guardian and the Economist, arguably two of the highest-quality newspapers in the world.

We worry that if LLMs can no longer train on quality data sources, training may be skewed toward fringe news sources who welcome the attention. If AI cannot access a premier scientific journal like Nature, will its training be overly weighted toward TikTok influencers, X, and Reddit? We risk **skewing AI-generated knowledge toward lower-quality, sensationalized, or ideologically extreme perspectives.**

And it gets worse. As AI models are increasingly trained on AI-generated content, cannibalizing it because high-quality content is restricted, we could enter a **self-referential loop** where AI starts **recycling and amplifying its own distortions**. This risks **a slow collapse in quality**, similar to how SEO-driven "content farms" polluted Google search results in the early 2010s.

Naturally, AI leaders have a riposte. One defense is that use of copyright materials falls under the **fair use doctrine** - a complex law that considers the purpose of the work, whether the work is fiction or non-fiction, the use's effect on the work's market value, and whether the usage is transformative. There's a legal gray area regarding whether such derivative works violate the IP rights of the original creators, particularly when the AI does not explicitly attribute or compensate the original creators. While human use of copyrighted materials in certain contexts (like criticism, comment, education, or news reporting) might be considered fair use, it's unclear how these principles apply to AI that uses such data for training and operation purposes.

Beyond fair use, the AI industry is increasingly focused on creating approved and authenticated datasets to address legal and ethical concerns surrounding copyrighted materials. Companies like OpenAI, Google, and Anthropic are exploring licensing agreements with content creators, media organizations, and data providers to ensure proper usage rights and fair compensation. Initiatives to develop "clean" datasets involve sourcing data from the public domain, purchasing licenses, or using synthetically generated data to avoid copyright disputes. These efforts aim to establish transparent and legally compliant data practices, mitigate risks of IP infringement, and bolster trust among creators, regulators, and users. These authenticated datasets are sure to become part of the AI landscape, but they raise a tricky question – **who gets to say what is authentic?** We have friends that only read scientific journals and The New Yorker, and we have friends for whom any information from the mainstream is corrupt, preferring X and Reddit.

In a twist on the intellectual property debate, early 2025 saw a new player explode onto the global AI scene. Previously obscure Chinese startup DeepSeek launched their chatbot in January, rapidly overtaking ChatGPT as the most popular free app in Apple's app store. DeepSeek made bold claims about the speed and size of their model, which appeared to have been developed at a fraction of the cost of other comparable LLMs. This stunning development roiled markets, with investors forced to reassess their assumptions about what it takes to build and train a frontier model.

With deep irony given their previously blasé attitude to charges of copyright infringement from artists and authors, OpenAI executives were publicly furious at what they claimed was DeepSeek's illegal use of their model's data for training purposes. The general consensus among AI users and commentators: karma.

In our view, the years 2025 to 2030 will see open warfare (if we may be hyperbolic) between content creators and proliferating LLMs (and LMMs and other transformer species). We find ourselves unsure of which side to land – high-quality content, from newspapers to artists, should be incentivized and protected. Yet, if our intuitions about the usefulness of AI to humanity are correct, we should also worry about the dumbing down of LLMs.

What sets these three issues, opacity, emergence, and intellectual property apart is that you and we can do nothing at all about them.

On these issues, we lack agency. The decisions about how AI was developed, and how fast, and how it was trained were made in Silicon Valley's culture of **permissionless innovation** (which we get to later). Likewise, the decision about whether humanity should pursue AGI and ASI as a goal was made by a very few people, many of whom with noble intent, but seldom very far removed from the profit motive.

Robustness and safety

Robustness is the ability of a biological or technological system to maintain its function despite environmental changes or stress. As we depend more on AI, robustness is key: think a new malware attack.

But systems can be robust without being safe. In real-world applications, unpredictability is the norm, A self-driving car could perform under adversarial conditions, but fail to avoid pedestrians. And systems can be safe without being robust; that is, safe under normal conditions, but dangerous under stresses. Lack of robustness in AI is a function of emergence and unpredictability we saw earlier. Design In healthcare, finance, or autonomous vehicle, where human safety may be affected, should include standards for **fault tolerance, edge testing**, or the system's ability to handle **extreme loads.** But **opacity**, inability to understand the system's logic confounds "debugging" when it breaks down.

Unpredictable results, because of emergence, may be easy to spot when it comes to counting the Rs in strawberry, but on problems humans can't easily answer, we may not spot an error. Because the internal logic is opaque, we may not be able to fix it either.

Robustness and safety cost money and business leaders have often fought regulation or internal controls on robustness and safety because they can cost money. In the textbook example, the Ford Pinto was prone to explode in rear-end collisions because of fuel tank positioning, but in a now-infamous cost-benefit analysis, Ford concluded that it was cheaper to settle lawsuits for injuries and deaths than to fix the fuel tank design. The estimated cost of redesigning the car was about **$11 per vehicle**, while potential liability costs were calculated to be less. Is that all a life is worth?

At the Deepwater Horizon rig, engineers faced a decision between a swift installation of 6 centralizers to get the rig up and running or a delayed installation of the 21 centralizers recommended by computer safety modeling. Costs on the rig were accumulating at $1m a day, so 6 it was. Boom! $65 billion in fines, fees, and costs, not to mention an environmental disaster.

Leading AI firms are investing heavily in robustness and safety measures, often collaborating through cross-industry working groups like Partnership on AI and MLCommons, and engaging with non-profits such as The Future of Life Institute and The Center for AI Safety. These efforts focus on developing standards, conducting safety research, and sharing best practices to mitigate risks. Anthropic, in particular, is taking a unique approach with its "Constitutional AI" methodology, which involves training AI systems with a set of guiding principles (a "constitution") to instill safe and ethical behavior from the ground up, rather than relying solely on post-training reinforcement. Time will tell whether such efforts were sufficient and timely, or not.

When an AI system fails, is not robust, we want **accountability**. That is easier said than done with AI. If we can't predict what AI will do, who is accountable when it doesn't do what we expect?

Accountability

On a Friday afternoon in July, your Australia-based author was cramming a few last tasks into the working week when he got the dreaded blue screen of death. From the groans around the office, so had everyone else, so they did what Australians like to do and headed to the pub. There, they discovered that not only were office computers down, but that they were in the middle of the largest outage in IT history with 85 million computers crashing around the globe.

The carnage spread to banks, supermarkets, and airlines. The cause of all this chaos? A faulty update from AI cybersecurity provider CrowdStrike. One cyber risk consultancy, Kovrr, estimated a cost to the UK economy of $2-3 billion. Delta Airlines alone claimed $550 million in losses and sued. Malaysia's giant, AirAsia, followed suit.

Who is to blame when AI goes wrong?

If a self-driving Tesla takes a dip in a pool (see Figure VII.5) or kills a pedestrian, who is responsible? With AI denied legal personhood (for now), liability falls on someone else, but who? Is it the developer of the AI model, the providers of the AI's training data, Tesla leadership, or the passenger who outsourced driving to the car's AI? Tesla writes its own code, but most companies buy or borrow AI components, complicating the accountability chain. Should we look to the seller of the car, the algorithm or data supplier, the marketing team promising specific outcomes, or the government for insufficient regulation?

When individuals use AI for advice or research, the accountability lies with them, as most platforms warn: *"AI can make mistakes. Check important info."* But what happens when the stakes are high? It's all very well to state that accountability should remain with the human user. However, the convenience of AI, not to mention the

FIGURE VII.5: LA summers are hot, but seriously?

veneer of intelligence, quickly tempts users to "set and forget," by-passing essential human oversight. (Recall our earlier example of a zealous lawyer issuing a brief with fictitious case law generated by AI.)

Leaders face a dilemma. Encouraging employees to explore AI's potential fosters innovation and productivity. However, unregulated experimentation introduces significant risks. Too many restrictions stifle creativity; too few jeopardize the business.

Leaders must adopt a "belt and suspenders" approach to accountability. The "belt" is human-in-the-loop (HITL) systems, where AI augments human judgment rather than replacing it. HITL requires formalizing authority, oversight timing, decision review processes, override mechanisms, and error correction protocols. For high-stakes environments like law or medicine, this means structured processes rather than informal reviews.

The "suspenders" include transparent algorithms, continuous learning mechanisms for AI, and "red-teaming" (see the last chapter in this section) to identify vulnerabilities. Leaders also need to ensure fluency in ethical principles across all levels of their organizations. Training employees with real-world ethical dilemmas prepares them to navigate tight deadlines, demanding clients, and other pressures while upholding accountability.

Finally, accountability must exceed liability. Ethical accountability involves addressing harm proactively, not just when legal risk is present. Firms that set their accountability boundaries at legal compliance risk falling short of broader ethical responsibilities, not least because the laws in this space already lag the technological capabilities and that gap is unlikely to close anytime soon.

We have suggested that AI's opacity, emergent behaviors, and potential for intellectual property theft present ethical challenges that demand thoughtful scrutiny. Yet, these "hardwired" features are only part of the story. Beyond the technical and philosophical complexities, real-world deployment of AI introduces pressing ethical issues that require our immediate attention.

In the next chapter, we explore five critical areas where ethical vigilance is essential: biases that can perpetuate injustice, privacy concerns in an era of pervasive data collection, cybersecurity risks that threaten both individuals and institutions, sustainability considerations as AI's resource demands grow, and, ultimately, the profound question of how AI can truly enhance human flourishing.

CHAPTER **VIII**

PEOPLE OR PROFIT?

In 1992, scholar Francis Fukuyama declared "the end of history." Liberal democracy and capitalism had won the day versus autocracy and communism and were apparently the culmination of humankind's ideological evolution. While the passing of one-third of a century has cast doubt on the liberal democracy half of that proposition, we are all some form of capitalist now. Yes, arguably even China. Very roughly, thirty-five to forty-five percent of China's GDP is from private enterprise. The US economy is about 70 percent private enterprise.[11] Fukuyama would have been more accurate to say that mixed-economy capitalism has prevailed, though that wouldn't have sold as many books.

Where countries vary widely is in government's role in directing capitalism and to what extent they offer safety nets for people who fall through the cracks. In broad brush strokes, governments can steer investment[12] toward riskier or long-term projects via **industrial policy** or attempt to rein in capitalism's "market failures" (environmental externalities, monopolies, exploitation, or financial crises) through regulation. The US mostly rejects industrial policy (strategic government investment in future technology and infrastructure) as a tool to guide investment. China, Singapore, and South Korea make much heavier use of it to steer innovation and build infrastructure.

In the **Intelligence Age,** AI offers unprecedented opportunities to enhance human well-being, advance science, drive economic growth, and address global challenges. Yet, as AI systems become

---

[11] These numbers vary depending on what is counted and how, so treat them as rough guides. China and the US aren't the extremes. Chad and North Korea have the least private enterprise, and Switzerland and Singapore rest alongside the US. Europe is in the middle with 45-55 percent of their economies private enterprise.

[12] There are famous exceptions, The New Deal, for example, but industrial policy is a dirty ("socialist") word in some circles. Despite this, the Biden administration did pass some. The CHIPS and Science act aimed at bolstering and reshoring the US semiconductor industry and several other similar acts were aimed at infrastructure and green energy.

increasingly embedded in business operations and society, we need to ask: Will AI be a tool for people or for profit? When workers are dislocated, as they will be, what kind of safety nets will be put in place? Is AI strategically important enough to merit vigorous industrial policy, as Singapore has decided? Should government attempt to guide investment toward human good, or allow the market to set the direction and tone of AI investment?

This chapter explores the tension between that capitalist backdrop and people-first ethics, examining whether AI development can genuinely align with humanistic, people-first ideals. The jury is out. But an idea called "stakeholder capitalism" (see below) is a partial answer.

Shareholder versus stakeholder capitalism

There is a strain of thought that under capitalism, when push comes to shove, when there are ethical trade-offs, capital always wins. Economist Milton Friedman was the poster boy for this view saying in 1971: "The only social responsibility of business is to make a profit." In the 1970s and 1980s, that **shareholder value capitalism** was the paradigmatic "true north" for business, according to austere institutions, such as the University of Chicago and the Harvard Business School. In this line of thinking, business ethics is the job of government, legislators, and the chief counsel's office. Business makes money, government makes laws. In this view, business ethics is, in math-lingo, an empty set. (See Figure VIII.1 for an irreverent take.)

Shareholder value capitalism, as a concept, has been relegated to the trash heap of business theory. But since capital wields vast political power, the idea lives on in practice, even if pointy-headed business school professors disdain it. The global leaders in AI, OpenAI, Google DeepMind, Microsoft, Nvidia, Meta, Amazon, Alibaba, Baidu, SenseTime, Anthropic, Apple, and Palantir are a Who's Who of Silicon Valley with approximately 20 trillion dollars in market capitalization. That amount of cash buys a lot of influence. Through lobbying and legal Jiu-Jitsu, "capital" doesn't have a thumb on the AI policy scale, it has parked its hefty great derrière.

"Somehow we need to monetise this - and quickly"

FIGURE VIII.1: Sometimes executives focus on the wrong things. (Cartoon by Steve Jones: jonesycartoons.com. Published by Prospect: prospectmagazine.co.uk)

Starting in the 1990s, a new "stakeholder capitalism" paradigm emerged. Managing a so-called **triple bottom line (TBL) of people, planet, and profit** not only was better for people and the planet but also made more money. How?

Many of those "people" in the 3 Ps were customers, communities, citizens, and employees. Treat them right and they come back. Environmental stewardship, the other P, reduced regulatory risks, brand damage, and resource consumption. An ethical no-brainer.

Research during this time emerged that companies who pursued this broader ethic were more profitable, enjoyed brand ad-

vantages, and led in the war for talent[13]. The zero-sum thinking of fifty years ago that ethically run businesses were disadvantaging shareholders proved naïve and empirically false. The tradeoff between capital, people, and environment wasn't zero-sum.

In the 2020s, in the US, the pendulum swung hard against this view with opinions that "triple-P" governance (contrary to the evidence) harms shareholders. This perspective depends on the assumption that what is good for shareholders is good for everybody: an {coughs} unlikely premise.

Much AI research and deployment is being done in the US, in the middle of a political and ideological tempest that makes it look like history isn't quite done with us: "the end of history" was a premature eulogy.

AI leaders are a smart bunch. They, perhaps better than we do, realize that AI is risky technology. And many have humanistic ideals, at least publicly. That has led to two major strands in **self-regulation policy**: **alignment** and **Responsible AI.**

Alignment: is it enough?

Ask many AI leaders about ethics and they quickly wave their sorcerer's wand: **Alignment**.

AI alignment is developing and training artificial intelligence systems in alignment with human **values**, **intentions**, and **goals**. In the simplest terms, WHAT are we going to teach it, and HOW are we going to teach it?

The first of those questions, what to teach it, is philosophical: **what** are human intentions, goals, and values? More pointedly, **whose** human intentions and values? Massachusetts'? Oklahoma's?

---

[13]   See Nogueira, E., Gomes, S. & Lopes, J.M. Unveiling triple bottom line's influence on business performance. Discover Sustainability **6**, 43 (2025.) It is one of dozen such studies, but the most current.

Denmark's? North Korea's? (How confident are you that Kim Jong Un wouldn't turn Seoul into a paperclip factory given half a chance?) And which goals: the UN's Sustainable Development Goals? Abstract goals such as human happiness or freedom? Economic goals? Longevity?

As a species, we agree on very little. Attempts to codify such values can lead to oversimplification or bias. We may get majoritarian (or average solutions) which have no guarantee of being ethical and which risk the exclusion of non-centrist perspectives. And what goals could we align upon that look reasonable to enough of our 8.1 billion people?

There's a temptation for alignment experts to stop at a short list of largely uncontroversial statements (akin to Asimov's "three rules of robotics[14]"). Some candidates are: don't harm children, respect privacy, don't inflict suffering, and don't deceive. But there are "hard cases" in each, for example, for the last two when a little suffering might prevent greater suffering or when a little deceit will prevent greater harm. Settling for an ethics that "every reasonable person would agree to" will produce too small a set of precepts, yet beyond the safe harbor of a small set of reasonable precepts lies a tempest of hotly contested values.

How humans decide what is right is the subject of 5,000 years of ethics writing. After all those years of philosophical study, ethics has defied boiling decisions down to a hard, exceptionless set of rules, condensing it into a "greatest good for the greatest number" calculus or even distilling it down to a supposedly universal set of values. Most often, ethical decisions require us to delicately weigh

---

[14]   For those of you who missed the movie, these are: 1) A robot may not injure a human being or, through inaction, allow a human being to come to harm. 2) A robot must obey the orders given it by human beings, except where such orders would conflict with the First Law. 3) A robot must protect its own existence as long as such protection does not conflict with the First or Second Laws. A pretty respectable stab for the 1940s!

different values, rules, and outcomes. In another classic thought problem, say a community of 100 can enslave one person and make 99 considerably better off. The "calculation" is easy, but our moral intuitions against slavery would prohibit it. In this example, two different ethical systems are at odds with one another – deontological (rule-based) ethics and utilitarian (calculated) ethics.

Anyone who has grappled with "trolley problems" will appreciate how hard this tradeoff makes ethical decisions.

Consider an example that initially seems trivial: A runaway **trolley** is heading toward a family of five stuck in a car on the tracks. You are standing next to a lever that will divert the trolley where it will hit one person instead of five. What do you do?

Easy game. You don't relish the choice, but ethics sometimes involves complex tradeoffs between lesser evils or greater good or between competing sets of values. You pull the lever.

Now imagine the same scenario where the trolley is heading toward the family, but to stop the trolley, you have to shove a portly individual off a bridge in front of the trolley. Most people recoil at this even though the "ethical math" is the same. Some values and rules must be factored in, in this case, prohibitions about murder.

Such apparently abstract thought experiments take on a different complexion with the advent of complex AI systems like those embedded in autonomous vehicles. Should one swerve to avoid the driver hitting a pedestrian but putting the driver, car, and passengers at risk? The developers of the software that powers self-driving vehicles (and the underwriters that insure them) have already had to grapple with the trolley problem and other thorny ethical dilemmas.

The computational models and reinforcement learning algorithms that underpin autonomous vehicle systems are so complex, and their logic so different from human reasoning that the trolley problem, tough though it is, manages to be less complex than reality. The kinds of ethical dilemmas produced by AI are many times more complex.

AI will create employment ethical dilemmas. Certain job class-es will be eliminated. Although AI hasn't dampened job markets in the West yet, as we approach 2030, the opportunity to down-size will be irresistible. In theory, in a "free market," customers get cheaper goods and services, and capital gets more efficiently deployed. The economic calculus is easy. But at the scale being contemplated, the slavery example from above should give us pause: Is 10 or 20 percent unemployment acceptable if the people who still work are better off?

What good is making the pie bigger if more people starve?

Layered on top of this is the awkward fact that corporate ethics is far more complex than a trolley problem or other thought experi-ment. Business ethics produces **wicked messes,** messes that add so-cial complexity because many different stakeholders with different needs, wants, values, and preferences are affected.

Even if we could wave our own wand and decide **what** to teach AI, **how** to teach it is a formidable programming challenge. Stuart Rus-sell from UC Berkeley and teams at organizations such as the Future of Humanity Institute at Oxford argue that traditional approaches of hard-coding rules or constraints are insufficient, as the above exam-ples demonstrate.

Instead, they propose approaches like inverse reinforcement learning, where AI systems attempt to infer human values by observ-ing human behavior. (That is a fascinating project, but is basing eth-ics on human behavior a good idea as it so often falls short of human ideals?)

OpenAI is working on RLHF (Reinforcement Learning from Hu-man Feedback), a machine learning technique where an AI model is trained using **human-labeled rewards and preferences** to refine its responses, ostensibly to align with human values. (Again, what are those?) Anthropic's alignment research team takes another ap-proach called Constitutional AI to attempt embedding ethical con-straints into large language models. Google DeepMind's safety team focuses on creating reward models that more accurately capture human intentions. These efforts include moral reasoning on what humans would want an AI to do after careful reflection and practical technical interventions like creating AI systems that can explain their reasoning, thus allowing a more transparent assessment of their alignment. (That sort of transparency is essential if humans are to understand the reasoning process and correct error.)

But transparency, as we saw, is tricky with the built-in opacity and emergence of AI reasoning.

Even the Nobel Prize-winning architects of AI systems cannot understand or explain, let alone predict, how their creations will act in specific situations. Many of the pioneers of AI, such as Herbert Simon and Geoffrey Hinton, believed that as they built AI, they would grasp the logic behind its intelligence. However, Hinton admits that this quest failed, and AI instead turned into an impenetrable black box.

The companies at work on alignment have some of the finest ethicists on their payrolls. Moreover, if we are to constrain AI's power or introduce additional safeguards or guardrails, it is these companies that will have to modify their behavior, perhaps at the cost of economic results. It is essential that they have skin in the game. But is that enough?

The next chapter tries to answer that question but first let's consider another self-regulatory framework: Responsible AI.

Is Responsible AI responsible enough?

"With great power comes great responsibility."
(Voltaire, but more famously Spiderman)

"Responsible AI" is the dominant voluntary ethical framework guiding ethical AI development and deployment.

It can be used in strategic planning and operations at the corporate level, but function heads should use it in their decision-making on how to use turnkey AI apps which are proliferating. For instance, marketing professionals need to understand ethical targeting, HR leaders need to avoid algorithmic bias, and finance heads need to get to grips with fair lending practices.

Today's most common frameworks have common features and some horrifying oversights. Figure VIII.2 compares four of the most prevalent.

Comparing Responsible AI Frameworks

Principle	OECD	EU	Microsoft	IBM
Human-Centric	Emphasizes human well-being	Priorizes human rights and dignity	Inclusive AI	Human-centered design philosophy
Transparency and Explainability	Emphasizes explainability and accessibility	Focus on transparency (e.g. disclousure of AI use)	InterpretML helps understand model logic	Watson OpenScale monitors models
Fairness	Emphasis on non-discrimination	Embedded in risk-based requirements	Core principle	Core principle
Robustness and Safety	Robustness is a key requirement	High-risk systems must meet strict robustness standards	Rehability and safety as a core pillar	Emphasis on system robustness
Privacy and Security	Included as part of ethical guidelines	Focus on data protection (GDRP integration)	Privacy and security are explicit pillars	Strong privacy commitments
Inclusiveness	Advocates inclusiveness	Ensures accessibility for all	Promotes inclusivity in design and deployment	Supports diverse and inclusive outcomes
Accountability	Requires clear accountability frameworks	Strict oversight via enforcement bodies and penalties	Accountability embedded in governance	Ethics board promotes accountability
Unique Features	Globality recognized, crosscountry alignment	Risk-based categorization, strict legal penalties (up to €30M)	Open-source tools (Fairlearn, interpretML)	Real-time bias tracking with OpenScale

© Paul Gibbons and James Healy, Adopting AI

FIGURE VIII.2: Responsible AI is a well-established standard for AI ethics, but is it responsible enough?

Each framework has a slightly different take. The OECD's has global reach and encompasses international standards; the EU's uses a risk-based classification and is aligned with the EU AI Act and its enforceable penalties. IBM's and Microsoft's business-oriented frameworks are similar but and include open-source tools for fairness, transparency, and real-time monitoring to operationalize abstract ideas such as fairness.

Notably and negligently, environmental sustainability and compassionate treatment of displaced workers are missing from these frameworks. They grotesquely neglect AI's systemic impacts, which are arguably just as important in a framework that could be called "responsible" in any meaningful way. It is as if two of the Ps in People, Planet, Profit have been partially discarded. A people-first governance framework content with the presence of some human issues, such as biases, but ignoring the effect of AI on jobs and employment seems callous.

Microsoft and IBM could counter that the function of governance is primarily to protect a for-profit enterprise, not humankind or the planet. Multi-national framers might argue that environmental and labor concerns fall under separate legislation (e.g., environmental policy and labor rights laws). Both could plausibly defend themselves from this critique by saying that such concerns are addressed elsewhere, by government or a separate policy framework.

Not good enough. We worry about the risks of AI ethics and governance frameworks sitting in a corner by themselves and not being integrated with mainstream governance policies. We worry that AI's effect on the environment and labor is too significant to have AI governance separate from labor and environmental policies. Why?

First, ignoring the societal consequences of AI undermines **public trust**. Will the public take seriously a business's commitment to responsible AI that conflicts with a company's stated environmental goals? Can a "human-centric" responsible AI framework ignore the human costs of displaced workers? How can you claim your framework prioritizes human rights and well-being (as they espouse) when omitting those risks omitting human costs from the ethical calculus around AI deployment? And already the Twitter-sphere (X-sphere), LinkedIn-sphere, Reddit-sphere, and BlueSky-sphere have begun to howl "BS" whenever the word responsible is trotted out.

Second, ignoring the other two Ps **squanders AI's potential**. AI has immense potential to combat climate change (e.g., predictive models for renewable energy, smarter resource allocation) and to help upskill workers through tailored education platforms. As with so much in the AI world, it can have huge positive or negative consequences depending on businesses' ethical commitments. If ethical governance is one lens through which AI strategy should be viewed, and if such strategies are to avoid harsh "externalities" (market failures), then the ethical lens through which they are viewed must consider **ethics in summa**, not just the narrow corporate ethics that gave way to better thinking in the last century.

And there are win-win-win scenarios where people, planet, and profit are weighed in appropriate proportions when setting strategy, which won't happen by zero-weighting two of those. It is tricky to find the strategic sweet spot, where profitability meets ethics, but leading companies find such win-win synergies, for instance, perhaps:

☛ Conducting **AI lifecycle assessments** to track environmental impact

☛ Setting **energy efficiency standards** for AI systems

☛ Establishing **reskilling funds** for workers displaced by automation

☛ Mandating **impact assessments** for AI deployments that affect labor markets or communities

Omitting environmental and societal dimensions in AI governance frameworks is not just blinkered, it undermines the credibility and efficacy of these frameworks. Responsible AI must recognize and act on AI's systemic footprint; otherwise, it is governance in name only, failing to deliver the ethical teeth it promises. Bridging this gap is essential for creating AI systems that truly benefit humanity and the planet.

Your authors like where Responsible AI started, but that was seven years ago, and like all gamers, are eager for the next patch to drop.

We understand companies will vary in their commitments to "triple-P" governance. Our project here is not to turn AI adopters green, although we would like that. Moreover, the zeitgeist is shifting away from fairness and inclusion. Like it or not, boards are retreating from such policies. Google has recently backed down on its prohibitions against military and surveillance use of its technology. Corporate values vary and change.

Moreover, different industries will have different AI ethical and risk profiles. For instance, i*n healthcare, AI governance must prioritize patient safety and data privacy; in manufacturing, the focus might lean toward workforce transitions and automation ethics.*

For this reason, **our strongest encouragement at this stage is for senior leadership to craft their own responsible AI framework** rather than importing one wholesale; a framework **guided** by well-established ones but tailored to their own values, market segments, and AI strategies. In doing so, leadership emerges with an up-to-date, fit for purpose, aligned ethical framework to guide their AI deployments.

Why? Over the years, we've seen hundreds of aspirational vision statements, values frameworks, and noble-sounding sustainability policies. Those are tepidly endorsed by senior leaders and rarely drive policy and behavior. We would rather have senior leaders craft a **proprietary responsible AI framework** for their business that they passionately endorse and enforce rather than pay lip-service to policies they are indifferent towards. What would be awful is if leaders used Responsible AI as a marketing veneer, as happened with sustainability greenwashing two decades ago (and still does).

Is self-regulation enough?

While there are obviously vital ethical questions to answer at a whole of a society, or even humankind level, individuals and organizations using AI day-to-day need to consider where their ethical boundaries lie. Outsourcing technology development to the broligarchs of Silicon Valley is one thing; outsourcing corporate ethics is quite another.

There is not yet, nor might there ever be, a society-wide discussion. Do we accept that the limits leading companies might place upon themselves are sufficiently stringent given the stakes of the game we are playing?

Helen Toner and Tasha McCauley, both former members of OpenAI's board, argue compellingly for a stronger role for government in regulating AI. Their essay, excerpted below, presents a case so well-reasoned that we felt it essential to include it here:

"Inside AI companies and throughout the larger community of researchers and engineers in the field, the high stakes—and large risks—of developing increasingly advanced AI are widely acknowledged. In Mr. Altman's own words, 'Successfully transitioning to a world with superintelligence is perhaps the most important—and hopeful, and scary—project in human history.' The level of concern expressed by many top AI scientists about the technology they themselves are building is well documented and very different from the optimistic attitudes of the programmers and network engineers who developed the early internet.

... in recent months, a rising chorus of voices—from Washington lawmakers to Silicon Valley investors—has advocated minimal government regulation of AI. Often, they draw parallels with the laissez-faire approach to the internet in the 1990s and the economic growth it spurred. However, this analogy is misleading.

It is far from clear that light-touch regulation of the internet has been an unalloyed good for society. Certainly, many successful tech businesses— and their investors—have benefited enormously from the lack of constraints on online commerce. It is less obvious that societies have struck the right balance when it comes to regulating to curb misinformation and disinformation on social media, child exploitation and human trafficking, and a growing youth mental-health crisis.

Can private companies pushing forward the frontier of a revolutionary new technology be expected to operate in the interests of both their shareholders and the wider world?

Now is the time for governmental bodies around the world to assert themselves. Only through a healthy balance of market forces and prudent regulation can we reliably ensure that AI's evolution truly benefits all of humanity."

Well, have they? In the scarcity of regulation globally, except notably in Europe, self-regulation and voluntary frameworks are the rule. What does the landscape of AI regulation look like globally?

CHAPTER **IX**

FIVE ETHICAL ISSUES
IN AI DEPLOYMENT

*It is not automation that destroys jobs, but the decision
to automate without concern for people.*
Douglas Rushkoff

In the blistering tech satire, *Silicon Valley*, startup founders compete in a mock "battlefield" pitching their ideas to seasoned VCs and executives. Each punctuates their pitch for funding with the same verbal tic: "...and we are making the world a better place." There has never been technology in our lifetimes that has attracted more optimistic rhetoric than AI, with evangelists waxing lyrical about the transformative potential for humanity and its effect on human flourishing.

Just how will that happen? We think that business has a critical role in scaling AI for the betterment of humankind, getting technologies from the lab bench to the living room, so to speak. If humanity is to prosper in the **intelligence transition,** business has an important role to play given that approximately 90% of spending on AI adoption comes from the corporate sector.

The five issues in this chapter are, unlike in previous chapters, issues over which business adopters have some agency. AI, in each of these areas, can be used for good, or for ill. AI introduces new cybersecurity vulnerabilities but can also strengthen cyber resilience. It can enhance human flourishing, eliminating drudgery and expanding innovation, or it can destroy jobs. It can augment surveillance or protect consumer and citizen privacy.

We first examine the most important of these, what AI may do to work and human flourishing, for work is where humans derive not just material benefits like a paycheck, but psychological benefits such as meaning and connection.

AI and human flourishing

"Work is about a search for daily meaning as well as daily bread, for recognition as well as cash, for astonishment rather than torpor; in short, for a sort of life rather than a Monday through Friday sort of dying."
(Studs Terkel, Working)

AI and jobs

To read the clickbait from the early 2020s: "**100 million jobs lost to AI and automation**!", we were all soon to be unemployed. The picture, as of 2025, is much less dramatic but more complex.

There is no real evidence (yet) that AI has affected employment, which throughout the early 2020s, particularly in the US, was the strongest in generations. Many firms surveyed expect AI to have a **positive** effect on headcount, particularly in the STEM professions, healthcare, finance, and creative industries. Nevertheless, labor economists expect that this will not continue and that certain roles will fare poorly: office support, customer service, and food services. (See Figure IX.1.)

Economists also predict that AI will add $15 trillion to the world economy by 2030, roughly another Europe. If accurate, that economic tailwind may stem the worst effects on employment. On the other hand, **a growing pie doesn't mean everyone gets more to eat.** During the outsourcing, offshoring, and automation boom in the 1990s and 2000s, manufacturing jobs in the US roughly halved. Americans got cheaper goods and corporate profits soared, but hardworking people and their communities in the Rust Belt suffered. Economists predict that AI will be just as disruptive, but unlike the 1990s, white-collar jobs will be affected, not just those in manufacturing.

The economic tailwind of the post-Covid 2020s won't last forever and optimistic cries of "augment, don't automate" won't stem the coming job losses. Millions of people will likely lose their jobs. How businesses do that is an ethical issue.

Potential AI Impact on Different Occupations

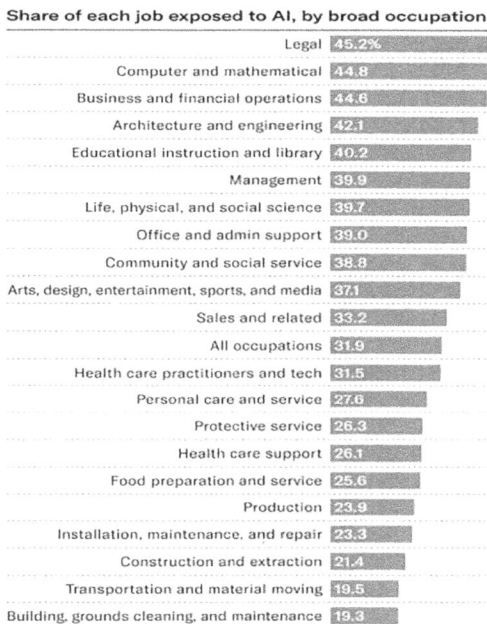

Share of each job exposed to AI, by broad occupation

Occupation	Share
Legal	45.2%
Computer and mathematical	44.8
Business and financial operations	44.6
Architecture and engineering	42.1
Educational instruction and library	40.2
Management	39.9
Life, physical, and social science	39.7
Office and admin support	39.0
Community and social service	38.8
Arts, design, entertainment, sports, and media	37.1
Sales and related	33.2
All occupations	31.9
Health care practitioners and tech	31.5
Personal care and service	27.6
Protective service	26.3
Health care support	26.1
Food preparation and service	25.6
Production	23.9
Installation, maintenance, and repair	23.3
Construction and extraction	21.4
Transportation and material moving	19.5
Building, grounds cleaning, and maintenance	19.3

FIGURE IX.1: Some job classes will be affected much more than others by AI. (Source HBR, Fall 2024, Is Your Job AI Resilient? Shrier, D., Emanuel, J., and Harris, M.)

Doing downsizing right: a scalpel, not a chainsaw

Historically, plenty of business leaders have taken a callous stance toward worker livelihoods: Neutron Jack (Welch)'s and Chainsaw Al (Dunlap)'s leadership brands were slashing workforces, yet the former achieved mixed results and the latter disastrous. Gutting workforces sometimes boosts short-term earnings, but often at the expense of long-term value.

Rational and compassionate downsizing studies business processes for efficiency gains and their potential for automation. The playbook for doing this rationally (the scalpel) was written in the 20th century, but leaders sometimes toss the playbook out and issue *diktats* to all department heads to find 10 or 20 percent headcount reductions, slashing efficient and inefficient departments equally.

Sometimes, perversely, departments that add long-term value (such as R&D) and are treated as cost centers get axed the most. Study after study has shown that the chainsaw approach backfires, leading to long-term declines in stock performance and market share.[15]

The ethical questions for leaders are: if AI allows for a slimmer workforce, what is our responsibility to those whose livelihoods will be affected? Is that ethical responsibility limited to statutory minimum severance (where it exists)? How much should organizations invest in retraining and redeploying? Should they offer outplacement services (counseling, resume, and job search support)?

Retraining and redeploying workers has substantial business benefits, from morale and trust in leadership, to retaining tacit knowledge amassed over years, to relationship capital, to lower recruitment and onboarding costs, to creating a continuous learning culture, and to organizational continuity. But training and transitioning workers takes more considered effort than a pink slip. We have found time and again, even in firms that are listed among "best places to work" that knee-jerk downsizing is a "Wall Street-approved" way to drive shareholder value (although it's empirically doubtful that it drives long-term value).

Downsizing, while it must happen from time to time, is often lazy leadership and (from a humanitarian perspective) less ethical. It is our thesis that thinking about retraining and redeploying once the workforce reduction is on the cards is too late and indicative of leadership myopia. **The people-first approach is to future proof the workforce now, to support current AI adoption and future waves to come.**

Leaders will have to take an ethical as well as a commercial stand on this issue. Some firms never downsize, while others do at the drop of a hat (or the drop of an analyst's report).

_____

[15] See Steel, P., and House, A., "Short-term pain for long-term gain? A longitudinal meta-analysis of downsizing-financial performance relationships" Frontiers in Behavioral Economics, July 2023

AI and meaningful work

In 1900, 40 percent of the US workforce was needed to feed the population through agriculture and related industries; in 2020, it was 10 percent. A similar seismic shift affected a billion workers globally in the largest workforce disruption in history. Arguably, since agricultural work is tough and poorly paid, that 30 percent are better off, although one might wonder how many of those rural workers found their way into "the one percent" three generations later, or whether their descendants remain the "working poor," now in indoor low-wage, dangerous jobs. Many suggest that AI will dwarf the transition from agriculture to industry and then from industry to services.

Economist John Maynard Keynes predicted that by 2030 humans would be required to work fewer than fifteen hours a week. One of the great promises of the AI revolution is the tantalizing prospect that this might really happen. Even among those who agree with the prediction, there's fierce debate about whether this transition will lead to dystopia or utopia. The worry is that, though it sounds good, the baby gets tossed out with the bathwater; that is, in getting rid of work, we get rid of an important source of meaning and connection.

Meaningful work is central to human flourishing[16], and it is hard to predict the effect that AI will have. We are in speculative territory, but science fiction writers, philosophers, and economists have done some of the thinking for us.

The children's movie WALL-E presents one plausibly dystopian future, while remaining gloriously feel-good, which is quite a feat. In a plot reminiscent of an Elon Musk fever dream, Earth can no longer support life thanks to a combination of climate change and pollution, so plucky *Homo sapiens* escapes in massive spaceships where obese,

---

[16] Human flourishing is the 21st century replacement for ideas such as Maslow's hierarchy. It includes happiness, of course, but also meaning, achievement, relationships, and challenge. For why this "beyond happiness" framework matters, see *The Spirituality of Work and Leadership*, published in 2019.

supine humans ride around in anti-gravity cars sipping 48-ounce slushies and watching daytime TV (See Figure IX.2). The implication is that without work, humans lack purpose and become listless wastrels.

FIGURE IX.2: Is this what humans become when we no longer have to work or think?

Economists from right and left, and even Karl Marx, would have agreed. In Marx's most famous work *Das Kapital*, he celebrated **work as dignified and essential to our humanity**, with collective work seen as particularly noble. The idea of "conscious creation" was central to our imagination and creativity, what makes us human: "...a bee puts to shame many an architect in the construction of its cells. But what distinguishes the worst architect from the best of bees is that the architect builds the cell in his mind before he constructs it in reality." **Dignified**, **collective**, and **conscious** is a view of work that might well resonate with 21st-century workers.

Dignified or not, work is one of the central pillars of Western society. What Max Weber termed "the Protestant Work Ethic" is alive and well. It wasn't always so historically, and not every culture reveres labor as the West does; some value leisure, exploration, learning and family more highly than work. In today's work-centric culture, laziness is the ultimate character flaw shared unreservedly by both ends of the political spectrum; the right caricatures the lazy, undeserving poor, while the left decries the lazy, undeserving rich. Nineteenth century thinkers, poets, and nobles actually praised idleness which would seem absurd today.

So, what if there's no work? What if AI did all our thinking, imagining, and creating?

Marx assumed that technological progress would end the scourge of scarcity and render work irrelevant. Although he could scarcely have imagined a Large Language Model powered chatbot, perhaps he was right after all? In *The German Ideology*, Marx famously pondered what this post-scarcity world would look like, imagining a nirvana where we'd all be free to pursue whichever passions caught our fancy: "to hunt in the morning, fish in the afternoon, rear cattle in the evening, criticize after dinner".

Star Trek similarly assumes a **post-scarcity economy** where money is obsolete, and employment is no longer necessary for survival. The Federation relies on replicators (technology that can create anything from energy) and near-unlimited energy sources, eliminating material scarcity. Work becomes a choice, with individuals dedicating themselves to exploration, science, art, and improving society. Marx would approve; material abundance has eliminated exploitation, and work is for creative fulfillment rather than survival.

In a dystopian take, Blade Runner, advanced AI and replicants (bioengineered humans) perform much of the labor in society, but wealth is concentrated in corporations (particularly Tyrell Corporation that makes the replicants) while many humans live in poverty or migrate off-world for better opportunities. Marx would light his Manifesto on fire.

In some "post-work" futures, Universal Basic Income (UBI) is proposed as a solution; it will probably surprise most readers to learn that well-known scourge of socialism Richard Nixon was a fan as long ago as the 1970s. If it were politically possible, would UBI supplant our need for meaningful work, "a purpose not just a paycheck?" Perhaps not.

In a world where worker wellbeing is, or should be, a constant ethical issue for organizations, leaders must be alert to the dangers of exacerbating an already chronic situation. Our always-on society's addiction to smartphones has created a pervasive norm of 24/7 working. The cacophony of beeps, bleeps, and buzzes that soundtrack modern life are complemented by the overpowering over-stimulus of visual notifications breathlessly informing us of the latest email, SMS, WhatsApp, Telegram, Slack, or Teams message. With these **weapons of mass distraction** an integral part of modern work, it's no wonder many workers report feeling distracted, exhausted, and burned out.

A more cynical view suggests that many of the supposed benefits of the technological revolution are illusory. Is humanity's lot really bettered by the ability to play Candy Crush anytime, anywhere? Do Instagram influencers, YouTube cat videos, or TikTok really add to the sum of human flourishing? In a work context, how much value does the explosion in computer-related activity *really* add?

Is it possible that, far from conjuring a scarcity-free utopia of endless leisure, AI in fact creates a future so dystopian that workers reminisce fondly about the halcyon days of the 2020s when working hours were shorter, digital distractions were fewer, and burnout wasn't an epidemic?

Will AI give us more of what is annoying or injurious, or less? There is nothing baked into the technology that determines the answer. How it affects people is up to business leaders.

Augment or automate: AI and productivity

Fortunately, the zeitgeist in 2025 is to focus on augmenting workers rather than replacing them. Many new AI features will come from enhancements to software that workers are already fluent in. Email systems provide an option to write the first drafts of messages. Productivity applications create the first draft of a presentation based on a description. Financial software generates a prose description of the notable features in a financial report. Customer-relationship-management systems suggest ways to interact with customers.

These features could accelerate the productivity of every knowledge worker. (See Figure IX.3.)

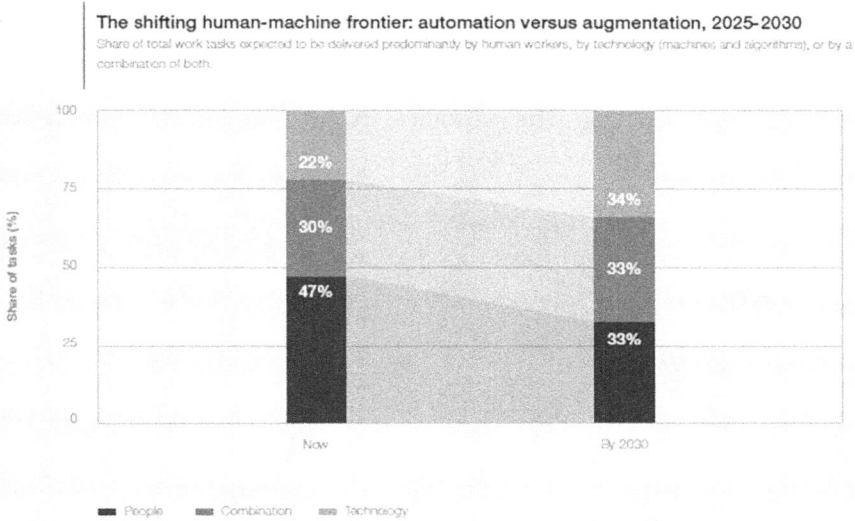

The shifting human-machine frontier: automation versus augmentation, 2025-2030
Share of total work tasks expected to be delivered predominantly by human workers, by technology (machines and algorithms), or by a combination of both

FIGURE IX.3: So far, few jobs have been lost, but predictions suggest that automation (job replacement) will accelerate. (Source WEF, Future of Jobs Report, 2025)

Nobel Prize-winning economist Robert Solow first identified the Productivity Paradox in 1987, observing, **"You can see the computer age everywhere but in the productivity statistics."** The power, sophistication, and usage of computers has soared since then, but productivity has continued to stagnate. Some attribute this to flawed measurement, claiming that economists' traditional concept of productivity simply can't capture many of the benefits provided by digital technology.

Support for Solow's perspective comes from an unexpected source. In a meteoric rise that uncannily coincides with the Generative AI explosion, the Four-Day Week movement has cut an unlikely swathe through the global working world. Evidence from their studies across the globe demonstrates that reducing the working week from five days to four, while paying workers for five days, has proved phenomenally successful. Revenue and profitability stay constant or, perversely, increase, while employee wellbeing increases as sick days and attrition fall.

The secret of this success isn't doing the same amount of work faster, it's forcing workers and leaders to make hard choices about priorities and take twenty percent of the work out of work. The great fear for AI is that it does the opposite. By automating so much, AI will create more time for workers to fill.

In some ideal world, probably one with unicorns and elves too, AI's benefits, costs, and risks are distributed equally. Do we care? Is there anything leaders can do?

AI and equality

The internet was supposed to level the playing field by democratizing access to knowledge, enabling startups and solopreneurs, creating new forms of social expression, and offering low-cost learning resources (e.g., Wikipedia and Khan Academy).

But by the late 1990s, we were instead talking about the "digital divide": uneven access to hardware and connectivity. Today, that divide is closing. Thankfully, 95% of US middle and high school students get a laptop, a trend accelerated by the pandemic, but it took decades.

As AI (rightly, because of its value to humanity) sucks in resources (people, capital, energy), how do we ensure humanity benefits equally (if we could even define what that means)? Can the AI divide be bridged more quickly than the digital divide?

We are sure, as sure as one can be with a prediction, that countries, companies, and workers who lead in AI today will out-compete and out-prosper their peers. Moreover, an AI advantage, as we have seen, is self-reinforcing (a virtuous cycle) as AI accelerates the development of more powerful AI but also the engineering infrastructure that enables it. We are likely heading for a **technopolar** world dominated by a few countries and a few companies with enormous economic and political power.

The top four AI firms, OpenAI, Microsoft, Amazon, and Google have nearly 86% of the market for models/platforms, and Nvidia has 90% of the data-center GPU market for training and running them. While larger firms can dedicate hundreds of millions to their "intelligence transformation," smaller firms will struggle to keep up, reducing their competitiveness. At the country level, the race is between the US and China, with the UK, EU, India, and Singapore trailing behind.

While the benefits of AI are unevenly distributed, the risks are not. This is called **moral hazard**: a few actors get the benefits of risk taking (profits, gargantuan salaries) while the downside is shared among the rest of us. The most telling example of this was the 2008 financial meltdown where the risk-takers enjoyed decades of profits and large bonuses, but when the excrement hit the fan, taxpayers picked up the tab.

A world where AI's benefits accrue to a very few people while the costs and risks accrue to everybody else would be unwelcome. Unfortunately, there's already widespread concern that AI-powered applications disadvantage some groups much more than others.

Biases

"Numbers have an authority that disguises their flaws."
(The Economist)

Porcha Woodruff was getting her two kids ready for school when there was a knock at the door. In front of her horrified children, six Detroit police officers arrested Porcha. She was charged with armed robbery and carjacking. At first, Porcha thought it was a prank: "Are you kidding, carjacking? Don't you see I'm eight months pregnant?"

It wasn't, and they couldn't. Or if they could, they didn't care. Because there was no doubt in the officers' minds: Porcha was guilty. They knew this for sure, because AI facial recognition software had matched Porcha's appearance to CCTV footage of the recent carjacking. Case closed.

After eleven hours in custody, Porcha Woodruff was released on $100,000 bail. Weeks later police dropped the charges for "lack of evidence." The carjacker in the video was obviously not pregnant and obviously not Porcha. But like Porcha, the carjacker was Black. That fact formed the centerpiece of Porcha's attempt to sue the Detroit police department for gross negligence.

This wasn't the first time that police had wrongfully arrested an innocent Black person based on faulty facial recognition. A Washington Post investigation found examples of fifteen US police departments arresting suspects purely on AI facial recognition[17]. Father-of-four Christopher Gatlin spent sixteen months in a Missouri jail after AI matched a grainy video of a hooded man in a surgical mask to Gatlin. Despite no ties to the crime scene or victim, no history of violence, and an alibi, police arrested and jailed Gatlin for a crime he could not have committed.

These cases share worrying aspects of the earlier COMPAS example (on opacity) from Chapter VII. They illustrate the dangers of reflexively relying on AI and the need for critical thinking and checks and balances to combat biases. A 2016 analysis by investigative journalists at ProPublica suggested that COMPAS was biased against Black defendants, **who were labeled as future criminals at almost twice the rate of white defendants.** Statistical evaluations revealed little difference between Black offenders and those from other groups, despite COMPAS predicting that Black offenders would re-offend twice as often.

Are human biases amplified by AI?

The problem with rooting out bias in AI systems lies in asking biased humans to design an unbiased system. The dictionary definition of bias, "inclination or prejudice against one person or group," sounds sensible but conceals a psychological, philosophical, systemic, and political quagmire.

---

[17] This is not just an American failing. While many other countries are more cautious than the US about facial recognition in law enforcement, there are many cases of wrongful arrest.

There are strong psychological reasons, rooted in human evolution, for bias. Gathering evidence, carefully weighing up options, and conducting rigorous logical analysis takes time and consumes lots of mental energy.

Imagine a group of your ancestors wandering the plains of east Africa. They're in an area known to be frequented by lions when they hear a rustle in the bushes. Most of them jump to an obvious and terrifying conclusion and hightail it. However, one of them, we'll call him "Logical Lenny," thinks that running based on one inconclusive data point would be hasty. Lenny sticks around to gather more data so he can accurately assess whether there's a lion. You don't need to be an evolutionary biologist to work out that Lenny only has to be wrong once before his overly logical genes are removed from the gene pool.

As a result, much human decision-making occurs automatically, below the level of conscious awareness. These automatic decisions tend to rely on heuristics, or simple rules of thumb, like "if there's a noise in the bushes in lion country, it's probably a lion." Rather than answering the often-difficult question of "what should I do?" in a situation, we tend to answer easier questions instead: "what's everyone else doing?", "what do I usually do?", or "what's the easiest thing to do?"

Most of the time this works just fine, but occasionally our tendency to make sense of the world using simple categories gets us into trouble. With just a CV and one interview to go on, it's hard to figure out if someone will be a good fit for a role, so humans fall back on rules of thumb, assumptions they may not be aware they're making, or biases they'd prefer they didn't have.

Human institutions have developed checks and balances to try to mitigate the impact of bias. Some organizations blind their hiring processes, and the jury selection process is designed to weed out obviously biased members. However, as we saw with COMPAS, and with the various cases of mistaken identity in Detroit and Missouri, there's a real danger that we assume AI is objective, consistent, and impossible to appeal.

The misidentification of Porcha Woodruff and Christopher Gatlin were examples of algorithmic bias, just one of the four types of AI biases leaders need to worry about: algorithmic, data, deployment, and development (shown in figure IX.4.)

Four Ways AI May be Biased

Source of Bias	Description	Example
Algorithmic	Biases in model architecture, feature weighting, or parameters.	A facial recognition system misidentifies indivuduals with darker skin tones due to insufficient attention to feature diversity during model design.
Data	Historical prejudices and underrepresentation of groups (sick, disable, very old or young, non-white, non-male).	Predictive hiring algorithms disadvantaging women because training data was sourced from historically male-dominated hiring patterns.
Deployment	How a system is deplyed and who gets to control it.	A credit-scoring AI disproportionately assigns lower scores in economically disadvantaged areas due to discriminatory deployment criteria.
Development	Which problems matter? Clean water in Uganda, or self-driving cars in LA? Who benefits?	AI investment prioritizing luxury goods recommendation systems over affordable healthcare diagnostic tools for underserved comunities.

© Paul Gibbons and James Healy, BOOK TITLE

FIGURE IX.4: Four sources of AI bias.

Data is the second source of bias. Today's LLMs train on human language, which can have gendered (and other) biases baked in. For example, the word pair MAN-WOMAN might lead to ACTOR-ACTRESS unproblematically. However, it might also lead to DOCTOR–NURSE or CEO-RECEPTIONIST.

We don't want our AI to believe that men **should** be doctors and women nurses. And given the data our AI trains on, it may be forgiven for thinking that way. If we trained an AI on boardroom pictures (especially from a few decades ago), it would probably conclude that company boards are predominantly white middle-aged men. A similar inference might be drawn from pictures of agricultural workers or other groups from a concentrated demographic. AI, as a baseball umpire would, "calls 'em as it sees 'em." But unlike an umpire, it doesn't understand the game's rules or context, leading to calls that perpetuate systemic inequities.

And perhaps worse, the inference about board composition could lead the AI to conclude that it *should* be that way. Because the AI lacks context, differentiating **is from ought to be,** is difficult. For example, if a position is historically more likely to be held by a man, should statistical filtering of resumes reflect that? Of course not.

These problems are not speculative, as Porcha Woodruff and others can testify. Amazon's AI hiring tool showed bias against women, forcing them to scrap years of development. Loan approval systems have been seen to disproportionately disfavor certain demographic groups.

Finally, there are deployment and development biases. In a nutshell, the most powerful AIs are owned by corporations and corporate investment is about 90% of the total currently invested in AI, dwarfing academic and government investment. That means problems such as securities trading and self-driving cars that look interesting commercially take precedence over curing malaria or solving climate change.

While bias is likely to remain a vexed issue for AI ethics, it's by no means the only one. AI can supercharge cyberattacks.

Cybersecurity and surveillance

"Yes, I'm paranoid, but am I paranoid enough?"
(William Gibson, science fiction writer, pioneer in the cyberpunk genre)

Deepfakes and information warfare

In February 2024, an accountant at British engineering giant Arup was summoned to a video call with the Group CFO and his team, to discuss an urgent and secret transaction. The employee initially suspected this was a phishing email but joined the call anyway and was quickly reassured by the familiar sight and sound of the CFO and senior leaders he knew. When the CFO ordered him to process a payment for HK$200m, approximately US$26m, he quickly complied.

There was just one problem. The CFO and every other person on the video call were AI-generated "deepfakes" and the hapless junior had just transferred $26m to scammers.

It's easy to scoff at the hapless Arup accountant duped by video, but a much more insidious version of the same scam targets ordinary individuals. Just three seconds of audio is enough for AI systems to realistically clone a voice, making it dangerously easy for scammers to create persuasive audio.

One common technique involves a call from an apparently traumatized loved one purporting to be in trouble with the police, or the victim of crime, needing money urgently. Others use realistic videos of celebrities or politicians to create fake ads: cryptocurrency from Elon Musk, or health supplements from Gwyneth Paltrow. Other techniques include romance scams and blackmail. All these rely on the toxic mix of human vulnerabilities and AI.

Deepfakes can be created using advanced machine learning tool called a Generative Adversarial Network (GAN.) They have legitimate uses, such as making Captain America appear to be 80 (see Figure IX.5,) but in bad actor hands they spread misinformation and disinformation, enable fraud and scams, and can be weaponized to create non-consensual explicit content.

FIGURE IX.5: Captain America looked pretty good at 80, but deepfakes have sinister uses in fraud and harassment.

Personalized fake news

Although fake news has been around since there was news, the Nazis began to systematically call newspapers *Lügenpresse* (lying press) when newspapers fact-checked Nazi propaganda. **Information warfare** creates and distributes disinformation to influence world events. A troll farm managing more than a hundred pro-Trump websites from an obscure Macedonian town called Veles targeted Hilary Clinton's presidential campaign in 2016. The young men orchestrating this scheme had no declared preferences in the US election, they simply found that Clinton-hate got the most clicks and the most advertizing revenue.

In 2016 Cambridge Analytica (CA) infamously harvested 87 million Facebook records to further the presidential campaign of candidate Trump and the Brexit referendum using "military disinformation campaigns and voter targeting."[18] The CEO-founder, Alexander Nix, even bragged about using such data in bribery sting

---

[18] From an interview with Nix on Channel 4 television, UK.

operations and "honey traps" as part of their "oppo research", a case of 21st-century technology augmenting the "oldest profession".

As well as manufactured fake news, such attacks are now AI-powered and often directed at corporations, flooding review platforms with negative feedback to undermine consumer confidence and amplifying negative stories to create public relations crises.

Consider that over half of adults get their news first from social media and that social media profits from clicks, not truth. AI supercharges content generation, valuable and harmful, and through sophisticated micro-targeting, makes sure misinformation lands in suitably fertile minds.

As the philosopher Lee McIntyre aptly notes: **"The truth isn't dying, it's being killed."** The twentieth century provided plenty of examples of" to "Our century has seen information being manipulated with generational consequences.

Supercharged cyberattacks

It's not just deepfake scams and fake news though. Other cybersecurity and surveillance threats enabled by AI include **phishing**, **privacy**, **verification**, **malware**, **vishing**, **data poisoning**, and **ransomware**.

AI gives hackers the ability to mount cyberattacks of huge scale and sophistication. Amazon chief security officer, CJ Moses, described the 2024 cybersecurity landscape this way: "On average, we're seeing 750 million attempts per day. Previously, we'd see about 100 million hits per day, and that number has grown to 750 million over six or seven months."

Increasingly, cybercriminals use AI to automate and power attacks, making them more sophisticated and difficult to detect. So-called **"script kiddies"** may lack sophisticated coding knowledge but AI turns them into master coders and powers their attacks despite lack of skills.

These attacks come in various flavors. Most cyberattacks, about 91% according to a survey by Cofense, begin with phishing emails.

AI-driven **spear phishing attacks** use natural language processing to create more convincing fraudulent messages tailored to the victim. Global phishing attacks rose by 34% in 2024, with roughly six million recorded (according to the Anti-Phishing Working Group.) Once data is stolen, AI tools make **privacy attack** amplification more effective by analyzing stolen data to extract information that makes messsages more targeted to the victim.

Human verification system attacks are testing cyber defenses. AI has long been able to decipher text-based CAPTCHAs, but if you've ever tried to select motorcycles from an array of blurry images while in a hurry, you may be relieved to know that those, too, have been broken by AI, making them nearly obsolete as well as irritating.

Catching **malware** is a never-ending game of whack-a-mole, but AI systems today can rapidly modify malware to evade detection and exploit new vulnerabilities. WormGPT emerged on the dark web in 2023 as an AI-powered malware tool that could generate extremely sophisticated phishing emails with a writing style far more engaging than the usual badly written Nigerian Prince emails.

AI also powers **adversarial attacks** where attackers manipulate model inputs (**data poisoning**) to further complicate the cybersecurity landscape. Moreover, integrating AI into critical infrastructure increases the potential impact of cyberattacks. If AI systems controlling essential services are breached, the consequences can be catastrophic, affecting public safety and national security. The rapid advancement and deployment of AI technologies often outpaces the development of corresponding security measures, leaving gaps that cybercriminals can exploit.

Finally, a Russian cybercriminal attempted to bribe a Tesla employee to plant malware for a **ransomware** attack; AI can empower such attacks by targeting vulnerable employees, such as those with vices who are susceptible to bribery. Finding new vulnerabilities in people, processes, or systems is how cybercriminals attempt to keep ahead of internal systems and controls. Google proved susceptible as recently as 2022 when a vulnerability allowed attackers to execute arbitrary code that exploited Chrome before Google could patch it.

Cyber security is a mountainous and complex topic, with a constant arms race between hackers and security professionals. Counterintuitively the effect of AI on cybersecurity could be net positive, making proactive firms more secure and forcing others to become more proactive.

AI fights back: cybersecurity defense

On the defensive side, AI tends to amplify whatever it is applied to, making both attacks and defenses more powerful. Accordingly, an early use case for AI was detecting and mitigating cyber threats. Because the landscape is evolving so quickly, AI's adaptability makes it much better than "hardwired" threat detection and mitigation, analyzing vast amounts of security data to identify emerging threats and attack vectors, so-called **threat intelligence,** enabled by products such as IBM's X-Force (among many others.)

Even better, machine learning can anticipate vulnerabilities: **predictive security,** a newish frontier in cybersecurity which can help businesses beat criminals at their own game. For example, "honeypots" are when a fake server entices an attack and gathers intelligence on the attacker. Predictive security uses **behavioral analysis** to identify deviations, such as unusual login locations, atypical file access patterns, or unexpected data transfers.

Once an attack is launched, AI empowers swift, real-time security responses, providing automated response capability. Moreover, AI's pattern recognition superpower enhances **anomaly detection.** Try using a credit card in a dodgy place or website, hypothetically, of course.

If you've ever wiggled your head around in front of your phone in a security check, you're familiar with **liveness detection**, also known as a **reverse Turing test**. These AI-powered authentication processes invert the traditional "Turing test" where a person tries to determine if another user is human or machine. The most sophisticated of these are "invisible" reverse Turing tests that monitor behavioral patterns such as scrolling, keystrokes, and clicks: AI validating the user's humanity without them knowing they're being validated.

This arms race between attackers, and defenders will be won by whichever side makes better use of AI. The same is true of our next topic, privacy.

Digital privacy

The cat was out of the privacy bag with the birth of the Internet in the 1990s. As citizens, we began unavoidably leaving a "digital exhaust" as we shopped online, posted and doom-scrolled social, used smart devices, or just read the news. This activity generates about 400 quintillion bytes of data daily, far more than humans could ever parse.

Then along came cloud computing to store all that data and powerful analytics to crunch it. Before analytics, banks couldn't possibly analyze one trillion annual credit card transactions, nor could social media companies dig into 140 billion WhatsApp messages, nor 50 billion daily social media posts. With analytics, your digital exhaust can be comprehensively analyzed in ways that may not be in your interest as a citizen.

AI takes high-powered analytics much further, enabling the mass collection and analysis of personal data, often without users' consent. Nobody ever signed anything that said, "Do what you like with my data, scrape away", although it is right there in the terms and conditions we all ignore. A 2018 academic study brilliantly titled "The Biggest Lie on the Internet" found that 74% of subjects didn't even pretend to read the privacy policy for a fictitious social network called NameDrop and 97% clicked accept, apparently oblivious to the clauses stating that all data would be shared with the NSA and the user's employer. It's as amusing as it is disturbing. Today, every time we shop, scroll, or swipe, we're all leaving digital breadcrumbs that Big Tech vacuums up, increasingly with AI.

AI's impact on privacy is complex and multi-faceted. There are eight data abuses made worse or easier by AI (see Figure IX.6.)

How AI worsens "data abuse"

Data abuse	AI effect
Data persistence	AI is trained on massive datasets; once data is distilled in statistical representations, it is impossible to remove – "immortalizing" it
Data repurposing	Machine learning thrives on data reuse and recombination, making repurposing easier and potentially remunerative
Data spillovers	Advanced inference can extrapolate information on non-target populations, perhaps including people who are not in a dataset
Data aggregation	Aggregation of data from multiple datasets leads to more intrusive profiling
Implicit consent	AI may vacuum up data without authorization. Even if citizens have explicitly given consent in the past (rare,) do they, by implication, consent to AI being trained on their data?
Anonymity	Datasets may be anonymized, but given enough data, AI may identify people leading to pseudonymity
Cross-border data	Data regulations vary by country, adding complexity to establishing training guardrails
Surveillance	AI enables advanced facial recognition, allowing tracking and profiling of individuals without consent

© Paul Gibbons and James Healy, Adopting AI (2025)

FIGURE IX.6: Data abuses worsened by AI

Big Brother is watching

In 2022, an invention called the mouse jiggler briefly took so-cial media by storm, allowing work-from-home (WFH) employees to move their mouse when they were AFK[19] to deceive surveillance programs installed by their employers. Alas, AI has improved worker

---

[19]   "Away From Keyboard"; a gamer thing.

surveillance so much it can now apparently tell the difference between a human hand and a jiggler device.

Your digital exhaust may also be surveilled without your permission through facial recognition software made possible by multi-modal AI, so-called Large Multimodal Models or LMMs. Scarily, you don't have to be online for digital surveillance. AI can identify you, track your behavior, identify you through your gait, and read lips! Various controversial initiatives use AI to interpret human emotions through facial expressions, a technology popular in the security industry, despite its reliance on the widely discredited (though still popular) theory of universal emotion.

On the other hand, ethical businesses can deploy AI to improve the security of consumer and citizen data. AI models can be taught to **anonymize**, not identify; AI bots can understand user **privacy preferences** and factor those in; AI-powered **cybersecurity** platforms can detect unusual data access patterns, strengthening protections; **"edge AI"** can keep data on user-protected devices; **advanced encryption**, enabled by AI, may harden security systems; AI can manage the **data lifecycle**, including the "right to be forgotten". As with each pillar, this is a classic crisis-opportunity for leaders and top firms are using **privacy by design** to upgrade their cyber security practices, derisk their businesses, and protect customers.

Aside from the vast increase in data collection and processing capability provided by AI, the major additional ethical wrinkle in this space is AI's opacity. This makes it hugely difficult for users or admins to see how private data is being used by an AI, increasing the likelihood of misuse, inadvertent or otherwise. It's imperative that organizations take steps to ensure their deployment of AI doesn't exacerbate the already fiendishly difficult challenge of ensuring data privacy.

While cybersecurity is fundamentally about protecting the business and its customers from digital threats, our next ethical issue asks us to broaden our view and protect something even bigger: the planet itself. As we pivot to the topic of sustainability, we shift our focus from securing data to securing our shared future, exploring how AI can be both a culprit in environmental harm and a catalyst for global change.

Environmental sustainability

Business makes sustainability commitments for three reasons: efficiency, branding, and innovation.

Efficiency is the no-brainer as sustainability commitments, say, around energy or resource use, often make great economic sense. Some firms famously bake sustainability into their **brand**, such as Patagonia and The Body Shop, allowing the former to charge $700 for a jacket. For other firms, environmental sustainability is a driver of **innovation**, and central to product strategy; for instance, Toyota's Prius in the auto sector, and Unilever under the leadership of Paul Polman.

AI complicates sustainability commitments because it's an energy hog. Just training, not operating, an LLM generates as much CO_2 as five cars over their lifetime or nearly 1,000 cross-continent flights. "Inference", that is, operating, may generate much more carbon than that. This problem will get much worse before it gets better. Over the next 10 years, the WEF expects AI energy usage to increase about 10-fold.

Moreover, access to water is fast becoming one of the most important geopolitical issues of the twenty-first century. According to a recent study, "Google's data centers used 20 percent more water in 2022 than they did in 2021, and Microsoft's water use rose by 34 percent in the same period".[20] Given the sensitivity of the technology in data centers, the water used to cool them needs to be free of impurities; AI competes with humans for drinking water and irrigation.

It gets worse. Data center infrastructure and end user hardware are also being built at a rapid rate. Says the UN Environment Program,

---

[20] https://e360.yale.edu/features/artificial-intelligence-climate-energy-emissions

*"The electronics they [computers] house rely on a staggering amount of grist: making a **2 kg computer requires 800 kg** of raw materials. Microchips require rare earth metals, but mining those causes land degradation, water pollution, acid mine drainage, heavy metal contamination, local ecosystem destruction, and particulate matter emissions. Even as mining companies work to improve their environmental footprint, at the end of the use cycle is **disposal**: improper hardware disposal releases three of the most toxic metals to humans: mercury, cadmium, and lead."*

A few sustainability bright spots

It isn't all bad news. AI models are getting smaller and more efficient, as new kid on the block, DeepSeek, showed. Nvidia's new generation of Blackwell GPUs and data center accelerators will make inference more efficient. More importantly, there are many ways that AI might help us solve environmental problems: optimizing energy usage, renewable energy, smart grid, CCS (carbon capture and sequestration) technology, fusion, and geo-engineering.

Some techno-optimists predict that AI will revolutionize our relationship with the planet and solve global warming and biosphere degradation. Perhaps. Eventually. Even as this book goes to press, there have been AI-enabled breakthroughs in many of the above technologies. However, we'd have to enact what AI recommends and our track record of listening to experts on such matters is poor. If an AI said, "Stop digging fossil fuels out of the ground and burning them into the atmosphere," would Exxon say, "our bad, we'll get right on it"? It seems unlikely.

On the other hand, if AI can, for example, optimize renewable energy grids and accelerate fusion research, it could make a few of the sustainability worries disappear. The problem is time. With data centers being built at an incredible rate and the time lag between a fusion reactor prototype and widespread adoption perhaps fifty years, that's half a century that humanity does not have.

One explanation of the intention-action gap that bedevils the environmental movement is the sheer scale of the problem and the solutions required. It all just seems too hard: what can little old me do? With AI, there's a temptation for organizational leaders to default to similarly fatalistic sentiments about the environmental impacts.

Like the ethical issues driven by what AI *is*, however, it's important that organizations using AI don't simply delegate responsibility to the organizations building it. The nature of AI systems creates various somewhat intractable ethical issues. The question for organizations using AI is the extent to which those ethical issues shape the organizations' use cases.

The most socially responsible AI developers are conscious of the footprint their technology leaves. Satya Nadella has gone so far as to propose a new metric for measuring the efficiency of AI models: tokens per watt per dollar. In this way, we have a way of tracking not just the energy costs of intelligence, but perhaps a broader metric for understanding output in the intelligence economy. There are already moves to disclose CO_2 emissions up and down the supply chain, including those of suppliers and customers, but it remains to be seen whether this will affect AI energy disclosures.

As with all human problems, machine intelligence will not be what limits progress; human intelligence, our collective, cultural, and political will, will be what gets in the way.

The best outcome is that part of every business' AI strategy ameliorates its ethical profile and better protects stakeholders in each of the five areas; the second-best outcome might be a laser focus on the most important two or three for their business – not bettering their risk profile across the board but attacking the risks most relevant for their industry and business. The worst outcome is head-in-the-sand allowing digital pirates, cyber attackers, and ill-prepared workers or business units to drag the business into an ethical swamp and threaten the business' future.

Here are three of the dystopian scenarios introduced earlier:

- ☛ Scenario 1: **Breaches and backlash:** Without governance, AI systems exacerbate bias, leading to large-scale privacy breaches and causing public backlash, eroding trust, and slowing adoption across industries. Regulatory intervention becomes reactionary and stifles innovation.

- ☛ Scenario 2: **Unemployment and unrest:** Lacking labor transition policies, automation-driven job displacement triggers economic inequality, mass layoffs, and social unrest, creating resistance to AI deployment.

- ☛ Scenario 3: **Biosphere blowout**: AI models with high energy demands contribute to climate change. Without environmental standards, their unchecked proliferation worsens resource consumption and undermines global sustainability goals.

Consider for yourselves what probability you might assign to these. Our view? They are each closer to a Red/Black than a Triple-zero bet, more like 50-50 than 38-1.

So we need laws, but how do we legislate and regulate technologies evolving faster than our political and legal systems can keep pace?

The answers lie in understanding the rapidly developing world of AI law and regulation, where we'll examine how to balance innovation with responsibility, freedom with safety, and progress with equity.

CHAPTER X

AI LAW

Our moral responsibility is not to stop the future, but to shape it.
Alvin Toffler

A lthough few of us are lawyers or board directors, we all need to understand AI law. Recall Wells Fargo, Volkswagen, Enron, WorldCom, Cambridge Analytica, and Purdue Pharma. Not all those famous ethical breaches started with illegal board-level policies. The ethical breaches often came from middle- or lower-level employees. Workers broke the law because they didn't understand it, or because they thought management (implicitly) wanted them to. This implies that every employee should have a working idea of which AI laws affect their work.

Few foresaw that in the 2024 US elections, business regulation might prove a decisive issue; crypto, AI, DEI, privacy, and hate speech were among politicians' talking points. Business law is becoming everybody's business. As citizens, we need a clear view of what AI is law is so we know how well we are protected and whether we think that is sufficient. We may want to vote for political leaders who adopt the mix of AI regulation and innovation that aligns with our values.

The legal tail wags the ethics dog

To read some job profiles for a chief ethics officer, you might conclude that the job is all legal compliance and very little ethics. For many in that role, the main job is to keep the company out of legal jeopardy and ensure staff do not "steal the silver" (commit actionable breaches of ethics). The role is, in many companies: protect the company from the law and from bad actors inside and out. That approach could be called legalistic rather than ethical, the legal tail wagging the ethics dog.

The ethics mindset is, in contrast, **"Do less than the law permits and more than it requires."** We don't assess a person as ethical if they merely follow the law - we expect more of each other. It might be legal to cheat on a spouse, dump rubbish where no one sees you, persuade granny to part with her savings in a sketchy scheme, jack up prices of essential goods or drugs, or conceal unfavorable terms in legalese fine print. But those aren't ethically praiseworthy even though legal.

As humans, our conduct, when we are at our best, is guided by our moral compass. The law, for most people, is something they never brush with. Most people we know have never hired a lawyer to keep them out of jail, let alone surrounded themselves with a phalanx of them. The law is a backstop, a floor for ethical conduct.

Yet some businesses endorse the legalistic stance, walking up to the line drawn by the law and, if found on the wrong side, paying the penalties as the "cost of doing business." In one instance, a power plant decided that it was cheaper to pay emissions fines when caught than to install equipment to reduce emissions. The public should trust government to protect the rest of us from business malfeasance. But since business regulation generally lags business innovation, this "do what we can get away with" can be highly profitable, as Wall Street investment banks have proved time after time.

But ethics, properly drawn, is an inquiry into what is the **right thing to do**, not what one can get away with. Again, "we can do that, but should we?" is a question that should guide operational and strategic decisions at every level.

Business leaders must decide what their ethical "true North" will be: what hard ethical choices are we prepared to take irrespective of what the law may today permit? Given the bewildering speed with which AI technology is advancing, it seems naïve in the extreme to rely on regulation keeping pace. Just as it would be unwise for organizations to outsource AI ethics to AI companies, it would be equally imprudent to outsource AI ethics to regulators. Organizations need to decide their own ethical positions and bake them into the adoption of AI from the get-go.

The AI regulatory landscape

"Thou shalt not make a machine in the likeness of a human mind."
(The Orange Catholic Bible, The Dune saga)

We believe AI will "add another Europe" to the global economy in the coming years, a transformative boost equivalent to $20 trillion in economic output. This growth will accompany scientific advancements that make the 20th century appear sluggish by comparison. But alongside these opportunities lie existential risks and immediate dangers: discrimination, fraud, cyberattacks, safety concerns, environmental harm, and job displacement. Ignoring these risks in pursuit of the prize could lead to catastrophic consequences.

In Homer's *Odyssey*, the hero Odysseus must sail between the Scylla, a six-headed sea monster, and the Charybdis, an enormous whirlpool. He chooses to sail closer to the Scylla, losing a few sailors, rather than risk total destruction. Food for thought: perhaps our Charybdis is human extinction, as some foretell, and our Scylla is the loss of a few immediate commercial opportunities.

Governments face such a choice with legislating AI development and deployment. Do too little and face possible existential consequences and the ethical quagmire we outlined in the last chapter. Do too much and they risk business flight or stifling innovation. AI laws must strike a delicate balance between supporting technological innovation and addressing existential, ethical, legal, and economic risks.

Humans have, as of now, successfully navigated Odysseus' strait for **some** technologies. **Human cloning** and genetic engineering with CRISPR have been carefully regulated. Though it sometimes looked unlikely, we haven't yet blown ourselves to smithereens with **nuclear weapons** nor killed millions with **chemical and biological weapons.**

Our track record, though, is far from perfect. **Climate change**, driven by fossil fuel dependency, deforestation, and industrial practices, remain an existential threat, with global cooperation lagging far behind the urgency of the crisis. **Plastic pollution** strangles our ecosystems and infiltrates the food chain (they have been found in newborns). Cancer-causing and immune system-suppressing **PFAS** (polyfluoroalkyl substances or "forever chemicals") are in our water, soil, and bloodstreams to stay.

Unregulated innovation can lead to catastrophic outcomes, whether through financial crises fueled by under-regulated markets or environmental disasters caused by unchecked industrial activity.

Consider the 2008 financial crisis. Both your authors had front-row seats, one coaching investment banking executives during the most devastating economic meltdown in 70 years, the other observing the carnage from the derivatives trading floor at Credit Suisse. For decades, Wall Street resisted meaningful derivatives regulation, allowing unchecked financial innovation to flourish. When the house of cards inevitably collapsed, governments injected an estimated $15 to $20 trillion to avert what our friend, a bank CEO, described as a "Mad Max scenario." Despite these extraordinary measures, tens of millions of jobs were lost, and the global economy endured its sharpest contraction since World War II. This crisis serves as a stark reminder that innovation without guardrails can have devastating consequences.

Moral hazard, again, is when one party takes risks, and another pays the price. That is the risk we face today. Tech titans take risks, privatize the gains to shareholders (and pad their not insubstantial salaries) but when the chickens come home to roost, as they inevitably do, citizens pay the price. The price with AI, might be the costs of millions of displaced workers, or even greater, a rogue autonomous weapons system or other existential threat.

AI legislation—global comparisons

As of the first quarter of 2025, there are only two entities that have passed comprehensive AI legislation: the EU and South Korea. Given the transformative nature of AI and attendant risks, this scarcity of enforceable regulation is astonishing.

The playing field is shown in Figure X.1, a massive oversimplification that manages to still be complex.

Global AI Legislation 2025:
A Comparative Analysis

Region	Legislation/ Framework	Strengths	Weaknesses
European Union (EU)	EU AI Act	Comprehensive and ethically grounded Categorizes AI systems by risk levels, ensuring proportionate regulatory measures Prioritizes user rights, safety, and transparency Mandatory 3rd party assessments for high-risk systems Establishes high global standards	May be costly for SMEs to comply with May constrain innovation
United States (US)	No national legislation; light single-issue legislation in four states	Encourages startup and SME innovation	May make competing in global markets difficult Considerable moral hazard
China	Next Generation Artificial Intelligence Development Plan	Strong government support accelerates AI growth Recent guidelines emphasize human control and safety Includes sector-specific rules for generative AI and content moderation, focusing on data legitimacy and state values	Industrial policy, not legislation
South Korea	AI Basic Act	Consolidates AI-related regulations into a single framework, reducing fragmentation Emphasizes AI ethics and transparency, building trust in AI systems Includes measures to foster R&D, data access, and AI infrastructure development	Compliance costs and stringent requirements may disproportionately impact SMEs, especially in emerging industries
Rest of world	OECD AI Principles	Promotes global alignment and common principles across member countries Emphasizes inclusive growth, human-centered values, and transparency	Principles are non-binding, relying on member commitment Different countries may implement principles inconsistently Divergence in enforceability and implementation timelines

Note: The status and specifics of AI legislation are continually evolving. It's essential to consult official government publications and legal texts for the most current information.

© Paul Gibbons and James Healy, Adopting AI

FIGURE X.1: AI legislation could not be more fractured globally, from a regulatory vacuum to a comprehensive treatment.

Some countries, like Japan, Singapore, and China, have developed AI industrial policies to focus investment and innovation on AI and AI infrastructure. These frameworks often include ethical guidelines, but without compliance mechanisms, their effectiveness (from an ethical standpoint) remains in question.

The US, a leader in AI development alongside China and the UK, has struggled to pass meaningful AI legislation. There are non-governmental ethical advocacy groups, such as the Partnership on AI: a multi-stakeholder organization fosters ethical AI development, staffed by very impressive thinkers from academia, industry, and civil society. Without denigrating their exceptional thought leadership, the worm turns on whether NGO "fostering" is enough given the stakes we are playing for.

In 2022, the US government introduced the *Blueprint for an AI Bill of Rights*, intended as a framework to promote safety, bias reduction, privacy, transparency, and the ability to opt-out. However, without enforcement mechanisms, the Bill is more aspirational than actionable. Faced with legislative gridlock, the Biden administration issued an Executive Order urging safety, transparency, and trustworthiness, requiring federal agencies to appoint Chief AI Officers and conduct risk assessments. Even that skimpy Executive Order has been rescinded, and the Bill of Rights faces an uncertain future. Certain US states have enacted AI legislation, but these laws are vulnerable to federal "deregulation regulation" efforts. They will likely be overturned, effectively leaving the corporate sector to operate to self-regulate without government oversight. Sparse and toothless summarize the state of US AI legislation.

Meanwhile, the EU has taken a different approach with the *EU AI Act of 2024*. US AI leaders criticize it as overly restrictive, European businesses complain about compliance costs and reduced global competitiveness.

Yet history offers a familiar pattern: industries have repeatedly fought regulations that threatened their profits. The petrochemical industry resisted leaded gasoline bans, despite the long-established harms of lead as a neurotoxin. Automakers fought safety measures such as the air bag, and extractive industries opposed pollution controls. Smoking has long been the leading cause of preventable death, with the US Surgeon General citing it as such in 1964, yet cigarettes still kill nearly 500,000 people a year.

When faced with regulation, it is almost axiomatic that businesses will amplify objections through public campaigns and deploy armies of lobbyists and lawyers to weaken or stall enforcement. Their resistance, predictable, orchestrated, and extraordinarily well-funded, often prevails over rational analysis of harms.

The difference with AI is that the proximate concerns (e.g., data privacy, cybersecurity) are less immediately life-threatening, and the existential concerns (which go as far as extinction-level harm) seem remote, and to many citizens, not something they should worry about, nor could do anything about should they worry.

This resembles climate change. Today's disasters, such as heat deaths, wildfires, unlivable cities, and more extreme natural disasters, can be blamed on other factors. By definition, in a complex system like the weather, there are always multiple, interdependent causes. That gives climate deniers an easy out wrapped in a scientific veneer. Greater long-term threats (the Gulf Coast of America, the Netherlands, and Bangladesh becoming uninhabitable) seem too remote for voters to worry about and climate deniers can point to the fundamental uncertainty of predictive models.

Will that be the case with AI, large, long-term risks discounted because they are speculative and (anyway) we can't do much about them and short-term risks are somebody else's problem?

The EU gets tough. Tough enough? Far too tough?

"… during the Cold War, we obviously did not allow random start-ups to manufacture nuclear weapons in the nuclear corridor in Poughkeepsie."
(Marc Andreessen, venture capitalist)

The EU is significantly behind the US, China, and India; France is about 6th in AI innovation, according to the Associated Press. European research budgets are a fraction of those of the US and China and lack surrounding ecosystems for commercializing what leading research exists. Some observers blame regulation.

However, specific AI regulation, the EU AI Act came into force in August of 2024, too late for that to be true. EU critics can claim that its history with regulation made it likely that the AI Act stifled innovation long before it came into force.

Other observers blame the **regulatory culture** in the EU that follows the **"precautionary principle"**, ensuring technology is safe before scaling, while the US follows a **"permissionless innovation"** model, encouraging risk-taking and rapid scaling. Your authors would rather be a business innovator in a permissionless culture, but would rather be a citizen in a precautionary culture, where do you, dear reader, stand?

Yes, the European Union (EU) often has more stringent regulations than the United States in many areas, food safety, toxic chemicals, data privacy, plastic waste, car safety, and worker and consumer rights. The difference in regulatory culture, if that were the cause of Europe falling so far behind in AI, should have broadly affected technological innovation across the continent. This is not generally the case, and the EU keeps pace in many frontier technologies, such as nanotechnology, robotics, and biotechnology.

Whatever the complex causes, this may be the most significant strategic misstep for an entire continent in our time and perhaps in human history. (As Europeans both, we take no pleasure in saying this.) Unlike the US, where regulation lags far behind innovation, the EU has arguably done the opposite, putting the regulatory cart before the innovation horse.

The Act is far-reaching and by far the strictest in any jurisdiction. To soften the effect on business, the EU opted for a **risk-based approach** where legislation is tied to the risk of an activity, product, or technology rather than blanket prohibitions or restrictions.

The four risk levels with examples are shown in Figure X-2.

Risk levels in the EU's risk-based approach to AI regulation

Risk Level	Description	Example
Unacceptable Risk	Systems that manipulate human behavior exploit vulnerabilities, violate human rights, or perform social scoring and are prohibited.	**Real-Time Biometric Identification:** AI systems that perform real-time facial recognition in public spaces without consent.
High Risk	AI systems that pose significant risks to health, safety, or fundamental rights must meet criteris before deployed.	**AI in Law Enforcement:** Tools that assist in predicting criminal behavior, wich may lead to discriminatory practices.
Limited Risk	AI systems with specific transparency obligations — the user must be aware that they are interacting with AI.	**Deepfake Generators:** Tools that create synthetic media, requiring disclosure to prevent misinformation.
Minimal or No Risk	AI systems that pose little to no risk to users' rights or safety. These systems are largely exempt from additional regulatory requirements.	**AI-Powered Video Games:** Games that utilize AI to enhance user experience without impacting users' rights.

© Paul Gibbons and James Healy, BOOK TITLE

FIGURE X.2: Is a risk-based approach a robust way to regulate AI deployment?

How unreasonable is EU AI legislation?

Many of the most visible uses for AI thus far have been applied to consumers: algorithms that predict what we might buy or watch and use behavioral science to get us to click. This happens mostly below our radar and is mostly unobjectionable: your authors would prefer Amazon not offer them floral print dresses, or Netflix recommend they watch reruns of Paw Patrol. One could argue that these uses are manipulative but not harmful.

As AI deployment becomes more widespread, however, citizens and consumers may find themselves watched and manipulated in ways that aren't in their interests. Manipulations that are harmful are called **dark patterns** by behavioral scientists. While dark patterns are covered under the EU's AI Act, in the rest of the world, regulations are slimmer or non-existent.

The EU's highest risk category, "unacceptable" targets real-time facial recognition and dark pattern-type manipulation. That does not, to us, seem draconian. The second risk category, "high," includes systems where human health and well-being are on the line. The devil is in the details, but from a 10,000 meter perspective, those seem like good things to protect.

What did we learn from Europe's GDPR legislation?

When you open up a browser window and a banner appears at the bottom asking you which cookies can be stored on your computer, you are seeing the GDPR in action. US reactions to the AI Act mirror the reception of the EU's General Data Protection Regulation (GDPR), which was initially perceived as ultra-restrictive. Did the GDPR, as critics feared, stifle innovation and commerce? And what might the EU have gained from being a first mover in regulation?

Negatively, it can be argued that the GDPR slowed innovation and restricted commerce in the **short-term**. High compliance costs diverted a few resources from research and development, while ambiguities in enforcement may have led some companies to take overly cautious approaches, delaying data-driven innovations. Restrictions on data collection and usage hindered advancements in AI, personalized marketing, and other data-intensive fields. Commerce was initially disrupted, with some small businesses exiting the EU market due to high compliance costs, and digital publishers experienced reduced ad revenues from restrictions on third-party cookies.

Were the protections that EU citizens enjoyed as a result worth it when evaluated against these drawbacks? Hard to say, though there were **positive effects** of the legislation. Protection from harm is cognitively undervalued, citizens only notice protection when it isn't there, when real harm is done.

The GDPR acted as a catalyst for **privacy-focused innovation**, driving the development of technologies such as differential privacy, federated learning, and privacy-by-design frameworks. These innovations addressed regulatory requirements and set the stage for more ethical and sustainable data practices. It would be more accurate to say that the GDPR redirected some technology spend toward consumer protections rather than impaired it.

But the regulation fostered long-term benefits, including **global standardization** of privacy practices and increased **consumer trust** in businesses that embraced transparency. GDPR also positioned EU companies favorably in privacy-conscious markets, though critics argue it inadvertently bolstered Big Tech, as larger firms had the resources to comply and maintain dominance.

An overwhelming positive effect, from a European point of view, is that Europe gained a first-mover advantage in setting global standards for data protection. Similarly, the proposed AI Act aims to position the EU as a global leader in AI governance, and its risk-based model will likely become a template for many nations. Figure X.3 sets out some advantages and disadvantages of the EU AI Act.

The EU's AI Act seems a balanced approach to regulation, mitigating the harms of high-risk AI while fostering innovation in safer applications. By prioritizing ethical principles, the EU strengthens public trust and cements its leadership in responsible AI development. This model offers a blueprint for other jurisdictions seeking to harmonize innovation with societal safeguards.

The business case for AI regulation

What anti-regulation campaigners often miss is that there is a business case for AI regulation, for creating a trustworthy ecosystem that fosters innovation, enhances competitiveness, and mitigates risks. **Public trust** is essential for commercial viability in the long run because trust accelerates adoption. **Regulatory volatility** increases capital investment risk and many major AI deployments with longer

Potential Advantages and Disadvantages of the EU AI Act

Potential Advantages	Potential Challenges
Global Leadership: The EU's sophisticated framework will improve international AI standards, as GDPR did for data protection	**International Complexity:** Determining risk levels across the EU's 27 member state's corporations may be inconsistent or contested
Encouraging Innovation: Minimal regulation for low-risk AI applications allows startups to experiment and deploy systems quickly	**Innovation Trade-Offs:** Compliance costs consume resources and redirect product innovation
Public Trust: By addressing high-risk AI explicitly, the framework reassures stakeholders and encourages adoption	**Flight risk:** Restrictions on high-risk AI could lead to flight by companies who wish to pursue high-risk strategies
Competitive advantages: Attracts ethically-and safety-concerned customers	**Effect on SMEs:** May place burden on newer enterprises who lack internal mechanisms for oversight and implementation
Ethical Alignment: Promoting ethical AI use provides a competitive edge in global markets by reassuring clients and customers	
Scalability Across Sectors: A tiered approach enables flexibility and adaptation across diverse industries, fostering consistent regulation	

© Paul Gibbons and James Healy, Adopting AI (2025)

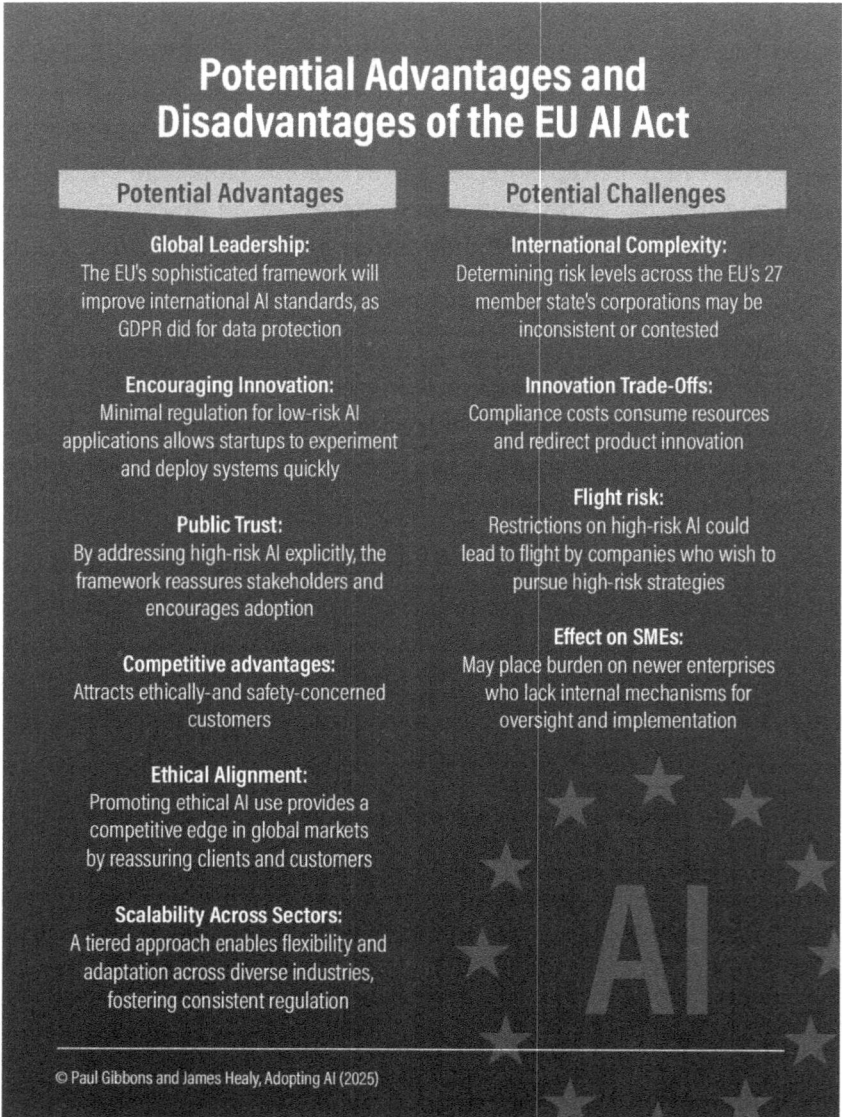

FIGURE X.3: Legislation is almost always viewed as a hindrance to businesses or countries. Is it always?

(say five-year) horizons will have a higher risk profile and a lower return on risk-adjusted capital. Businesses operating within regulated frameworks are better protected from **reputational harm**, and regulation also directs research and development in ways that the public would like it directed, towards **explainability**, **fairness**, and **accountability**, thus creating competitive advantages.

To shift the conversation around AI regulations, regulations must be reframed as guardrails, not roadblocks: regulation in the airline industry made it safer for passengers, increasing trust and supporting its explosive growth in the 1970s and 1980s. And air travel is unbelievably safe, given 100s of millions of people travel in aluminum cans five miles above the earth at nearly the speed of sound. A more robust regulatory environment domestically may **increase competitiveness internationally**; would you, as a purchaser, prefer an AI product developed somewhere with greater or lesser regulatory oversight?

We believe in the importance of what we have called the intelligence transition, convinced as we say in chapter I, of the overwhelming benefits to humanity of AI. Whether humanity's best interests are served by a regulatory vacuum in the world's leader in AI research is questionable. The problems we want AI to solve may take decades. Slightly slower deployment for much safer technology is a trade that we (speaking as citizens) would be prepared to make.

But corporations, whatever they may say, rarely play such a long game. With China's AI ascendance in 2024 and 2025, the fierce intercompany competition takes on an intercountry dimension that has national security implications for us all. Lax US policy may be defended in some circles by nationalist rhetoric about a Chinese threat.

This is what Dario Amodei fears is a race to the bottom. Nobody wants to get left behind, neither country nor company, and no government wants to be responsible for that happening.

Is the current state of AI Law "people-first?" Perhaps in Europe and countries following her lead, it is. Our principal challenge to business leaders is whether, if operating in a regulation-free zone, they should sail right up to the limits proscribed by the law, or follow their own "true North" compass in constraining some of their AI uses.

Businesses, we have said, should have their own Responsible AI framework, one that reflects their values, industry requirements, and one they will wholeheartedly enforce. That requires governance, the topic of the next chapter.

CHAPTER XI
AI GOVERNANCE

Corporate governance failures kill companies: Theranos' board lacked technical expertise and failed to independently audit their blood-testing product's claims; BP's Deepwater Horizon inferno released five million barrels of oil and cost them $65 billion in fines and had devastating ecological consequences; Volkswagen engineers installed a "defeat device" to deceive emissions tests, resulting in $30 billion in fines; Boeing's 737 MAX crashed twice, killing 346 people because profit pressure led to rushed development and insufficient testing, costing more than $20 billion and the public's trust.

Governance is like a car transmission: it connects policies, strategies, and ethics to actions and results. Without governance (the transmission), even the most powerful "engine" (strategy, policies, values) cannot propel a business forward. In today's fast-changing environment, governance functions are like modern adaptive transmissions, using sensors and software to respond to conditions like engine load and vehicle speed.

Governance for AI, with its unique risks and high stakes, demands similar adaptability and precision.

This brief chapter builds on the previous discussion of AI ethics and risks and explores how governance must evolve to address these challenges. We discuss AI Centers of Excellence, two AI governance roles, a set of AI governance tools, and sketch out a governance process bringing those together.

The governance gap

Businesses still lag far behind prudent AI governance. According to a Harvard Law School report, only 13% of large companies have a director with AI expertise, and only 0.8% have established an AI ethics board. Similarly, research from AIRS[21] suggests that only 40% of large firms have an agreed definition of ML/AI (which seems a sensible place to start), and only 10% have AI governance separate from

---

[21]   Artificial Intelligence/ Machine Learning Risk and Security

existing technology governance. That is atrocious given the technology's potential and risks.

There is no magic prescription for structuring AI governance. The choice between a Center of AI Excellence, a Chief AI Officer, or a Chief AI Ethics Officer depends on many factors, including the organization's size, strategy, culture, and existing governance structures. Additionally, organizations must assess whether these responsibilities can be integrated into existing roles, such as the CTO, CIO, Chief Risk Officer, or Chief Ethics Officer, or if they are best managed through board committees like Risk and Audit.

For some organizations, establishing a new role may bring needed focus and accountability, while for others, using existing structures may provide a more efficient and integrated approach to AI governance.

Should you create a Center of AI Excellence?

"Setting up a Center of Excellence is easy; making them excellent is hard."
(Paul Gibbons in Impact)

Yes, you probably should.

Centers of Excellence (CoEs) have become a familiar feature in many industries, supporting initiatives in areas like advanced manufacturing (GE), security (Microsoft), and sustainability (Unilever). Today, over one-third of Fortune 500 companies have established an AI CoE, reflecting the critical role artificial intelligence plays in transforming business practices.

COEs aim to promote knowledge diffusion in complex and emerging fields. COEs consolidate specialized knowledge and accelerate learning, acting as a repository for best practices and driving innovation. They encourage standardization and consistency, encouraging a cohesive approach across business units. However, the centralized nature of COEs can also lead to challenges. **Their effectiveness often relies on influencing decentralized business units** that may already have independent initiatives, sometimes resulting in resistance or poor adoption.

This presents a leadership dilemma. On the one hand, bottom-up innovation and technology adoption are to be welcomed. On the other, those "skunkworks" may neglect best practices or, sidestep ethical guidelines. When it comes to stakeholder touchpoints, say staff or customers, those local innovations may introduce business or ethical risks. Diffuse implementation also means business units don't learn from the successes and failures of others and organizational learning suffers.

Ironically, a poorly integrated CoE can create silos, isolating knowledge instead of disseminating it. Furthermore, while standardization fosters uniformity, it can stifle innovation by constraining experimentation.

Your authors have seen both successful CoEs and spectacularly ineffective ones, not because of a lack of technical expertise but because they lacked, for political or cultural reasons, the ability to influence critical business units. Often firms have paid big bucks for a name brand expert or advisor only to find that their advice fell on deaf ears. Our take? We would trade some technology skill for leadership and influencing skills in CoE big hires.

Power and effectiveness of the CoE

Centers of Excellence for AI governance vary in their approach to power and influence. Some operate with "soft power," offering guidance, expertise, and ethical advice without direct authority. Others wield "hard power," enforcing compliance with policies and ensuring adherence to ethical standards. The appropriate balance between these approaches depends on the organization's goals, culture, and the criticality of the AI systems involved.

In high-stakes areas like safety, fairness, and transparency, coercive power is often necessary. Non-compliance in these areas can lead to harm, making regulatory enforcement essential. For instance, ensuring ethical AI use, such as preventing bias or misuse, frequently requires mechanisms that go beyond voluntary adherence, addressing resistance or inertia driven by profit motives. However, relying solely on coercive measures can discourage innovation and create a compliance-driven rather than values-driven culture.

In less critical areas, voluntary frameworks and collaborative approaches may be more effective. These methods foster trust and innovation, particularly when mutual benefits exist, such as sharing data for responsible AI development. CoEs can play a pivotal role by providing expertise, establishing best practices, and creating frameworks while leaving enforcement to regulatory bodies. This balance allows CoEs to remain hubs of innovation and collaboration, focusing on enabling organizations rather than imposing restrictions.

Again, the success of an AI CoE depends less on technological brilliance and more on its ability to influence and integrate with existing structures.

Objectives of the CoE

An AI Center of Excellence could, if done right, be the strategic heartbeat of any organization's AI journey, setting the stage for ethical innovation and operational excellence. AI adoption is the most complex technical and multi-stakeholder challenge we have seen, reaching much further into an organization's DNA (culture, processes, strategy, technology) than, say, an ERM, CRM, or cloud adoption. So much so, that as we stressed in Chapter V, it's best thought of not as technology adoption at all.

Figure XI.1 shows a long menu of objectives for an AI Center of Excellence.

CoE objectives must be aligned with organizational goals, even if one of the CoE's important objectives is to reshape those goals. The AI CoE, unlike others we've created, doesn't just oversee AI development, it drives responsible innovation, empowers employees, and aligns AI initiatives with the organization's strategic vision.

AI Center of Excellence objectives

Governance and Ethics	• Establish AI governance frameworks and policies, ensuring responsible AI development and deployment • Develop ethical AI guidelines addressing fairness, transparency, accountability, and societal impacts • Implement comprehensive risk management frameworks, including monitoring for bias, model drift, and performance issues • Design data governance strategies with quality standards, data lineage tracking, and master data management • Develop incident response and business continuity plans for AI-specific risks
Standardization	• Develop standardized methodologies for AI project implementation, including problem identification and testing protocols • Establish documentation standards and knowledge management systems to reduce redundancy and transfer knowledge • Create processes for model validation, testing, and production monitoring • Develop vendor management and evaluation frameworks for AI tools and platforms
Talent and Capability Development	• Create and maintain an AI talent development program with training, skill paths, and knowledge-sharing initiatives • Build external partnerships with academic institutions, research bodies, and industry organizations
Infrastructure and Technology	• Build and manage a centralized AI infrastructure for efficient, scalable, and secure development and deployment • Foster innovation through sandbox environments, PoC frameworks, and pathways to production for successful pilots • Implement tools for monitoring model performance, A/B testing, and automated systems to track benchmarks
Strategic Alignment	• Align AI initiatives with business objectives through evaluation criteria, project portfolios, and ROI tracking • Establish partnerships with business units to identify experimental uses and facilitate adoption
Collaboration and Integration	• Create cross-functional collaboration mechanisms between data scientists, domain experts, and business stakeholders • Drive adoption of AI solutions through stakeholder engagement and organizational support
Metrics and Reporting	• Develop metrics to track AI initiative success, including technical performance, business impact, and adoption rates • Measure and report on the ROI and organizational value of AI projects

© Paul Gibbons and James Healy, Adopting AI (2025)

FIGURE XI.1: AI Centers of Excellence have a lot to consider, but can drive innovation and support enterprise-wide governance.

That leaves the question of who the CoE reports to. In many businesses, it may report through the CTO, but if it is going to oversee development in part of its role, that may not work. Further, reporting through the CTO may deemphasize the strategic, people, and culture aspects of the role. In our view, the strategic and ethical importance and broad remit require CEO sponsorship. Only with that level of sponsorship can the CoE integrate with the rest of the business, without which it will underperform.

Should you hire a CAIO?

The decision to establish a Chief AI Officer (CAIO) role reporting directly to the CEO depends on the strategic importance of AI to your company. They don't come cheap, between $350k and $1m in total compensation, according to Heidrick & Struggles. When should the board consider one?

Most organizations still assign AI oversight to existing roles, such as Chief Technology Officers (CTOs) or Chief Information Officers (CIOs). A 2024 survey found that 54% of organizations had a head of AI or AI leader, but only 12% specifically designated a CAIO to manage the overall enterprise AI strategy. The problem is, as we've seen over our careers, IT can be just terrible at engaging with "the business." AI strategy and culture-building can be secondary priorities compared to broader IT and operational responsibilities.

A CAIO can provide the leadership and focus needed to integrate AI into the company's strategy, operations, and workforce. They will support innovation, facilitate strategy development, help the board set ethical standards, and support AI deployments. Beyond measurable value, a CAIO reporting to the CEO signals to stakeholders, investors, and employees that the company views AI as a critical growth driver, elevating its strategic priority.

However, the value of a CAIO depends on the organizational structure and existing leadership. If responsibilities for AI are already distributed effectively among key executives (e.g., CTO, CIO, or COO), adding a CAIO might create redundancies, political strife, or siloed decision-making. In these cases, embedding AI governance and strategy within existing roles could be more effective than creating a standalone position. The key is to ensure that AI initiatives have executive-level visibility and sufficient resources to thrive, whether through a CAIO or by integrating AI leadership into existing structures.

The CAIO role works best when the company is at a maturity level where AI leadership requires a dedicated executive to navigate its complexities. This includes overseeing ethical governance, scaling AI systems, and ensuring ROI on AI investments. If AI is still exploratory, a CoE or an existing executive championing AI may suffice. But for companies betting on AI to define their future, a CAIO can be the visionary and executor needed to ensure success.

What about a CAIEO?

Several prominent technology firms, including BCG, Salesforce, IBM, and Microsoft, have appointed roles dedicated to AI ethics, such as Head of Responsible AI, AI Ethics Global Leader, Global AI Ethicist, or Chief Responsible AI Officer. The role is complex, requiring a background in ethics, technology, and business alongside superb communications skills.

As of today, only the very largest AI-native companies have such a role. As an alternative, some companies form an AI Ethics board with cross-functional representation, from technology, legal, risk management, HR, and marketing.

Creating a Chief AI Ethics Officer (CAIEO) role reporting directly to the CEO depends on the complexity and ethical stakes of a company's AI initiatives. If your organization deploys AI in high-stakes contexts, such as healthcare, finance, or law enforcement, where ethical lapses could lead to significant harm or reputational damage, a CAIEO can provide dedicated leadership to address these challenges.

The necessity of this role also depends on how your organization currently manages AI ethics. If ethical oversight is fragmented or relegated to middle management, appointing a CAIEO can centralize and elevate this critical responsibility, ensuring it receives the visibility and resources it requires. A CAIEO can act as a bridge between technical teams, legal counsel, and leadership, establishing clear frameworks for fairness, transparency, and accountability.

Reporting directly to the CEO emphasizes the importance of ethical AI use at the highest levels and reinforces trust among stakeholders, including customers, regulators, and employees.

For companies deploying AI at scale or in sensitive areas, though, a CAIEO provides the focus and authority to manage the ethical challenges AI presents.

Introduction to AI Governance Tools

With the surge in AI deployment, AI governance also requires some new tools, from software tools to auditing tools to governance training. There are two classes of tools that we offer a few examples of in Figures XI-2 and XI-3: software tools, some from technology leaders and some from smaller startups, and low-tech "analog" tools such as risk registers and red-teaming.

With newer software tools, and repurposed analog tools, AI governance is becoming easier. However, recall the governance gap that opened this chapter: just over 1 in 10 companies have a director with AI expertise.

The gap, as so often with AI, is a people gap, not a technology gap.

The AI governance process

The list of CoE objectives above paints a picture of a taxing and important governance agenda. Businesses scaling their governance function will wonder where to start, and although there is no one-size-fits-all process, we've broken down the governance process into four steps and four categories. Firms will need to build their own and should consider this illustrative rather than definitive. (See Figure XI.4.)

Governance is too often viewed as something rarefied for the board to worry about. However, a cursory examination of every ethical failure reveals that ethical breaches happen in the engine room, and not just at board level. Consider the Deepwater Horizon explosion: the

Software tools for AI governance

Purpose	Example	Description
Explainability and Interpretability	Local Interpretable Model-agnostic Explanations (LIME), SHapley Additive exPlanations (SHAP)	Enhance transparency by making AI decisions understandable to stakeholders
Bias Detection and Mitigation	Aequitas, Fairlearn	Identify and address biases in datasets and models to promote fairness and inclusiveness
Model Risk Management	IBM Watson Studio Model Risk Management	Assess and monitor risk associated with AI models, encouraging safe, reliable, and intended operation
Ethical AI Assessment	Assessment List for Trustworthy AI (ALTAI)	Evaluate AI systems against ethical standards like fairness, accountability, and transparency
Data Privacy Management	Immuta Privitar	Ensure compliance with data protection laws (e.g., GDPR) and safeguard user privacy in AI systems
Policy and Compliance	IBM's AI FactSheets, Google's Model Cards	Track model lineage, standards compliance, and adherence to organizational AI policies
Model Inventory and Management	ModelOp, Enzai's Model Inventory	Catalog and monitor AI models, ensuring proper documentation, version control, and governance
Governance Workflow Automation	Zapier, Microsoft Power Automate	Automate governance processes audits, risk assessments, and compliance checks
Security and Privacy	Google SAIF (Secure AI Framework) Microsoft Security for AI	Ensure secure handling of data within AI models while protecting sensitive information

© Paul Gibbons and James Healy, Adopting AI (2025)

FIGURE XI.2: Although too few organizations have dedicated AI governance roles, there are dozens of software tools.

fatal decisions were made by engineers at the wellhead. At Volkswagen, the "defeat device" was installed by engineers on the shop floor.

Governance cannot remain confined to executive committees. Organizations can empower managers at all levels by providing tailored AI ethics training, equipping them with decision-support tools

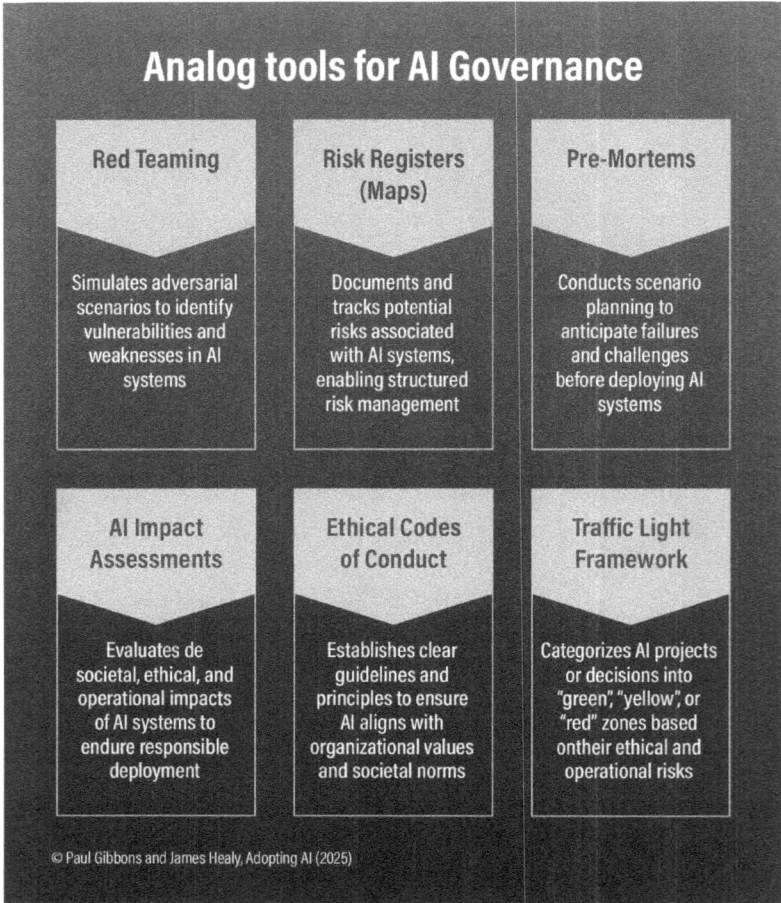

FIGURE XI.3: Analog tools for risk management and ethics adapted for AI governance.

like ethical checklists, and integrating AI governance criteria into performance evaluations.

We recommend cascading AI governance as part of mandatory ethics training. Such an education might cover ethics, compliance, regulation and the latest developments in the field. Users should understand risks, potential harms, privacy, deepfakes, AI limitations, copyright and much more. (Our AI ethics framework provides an excellent basis for such a course.)

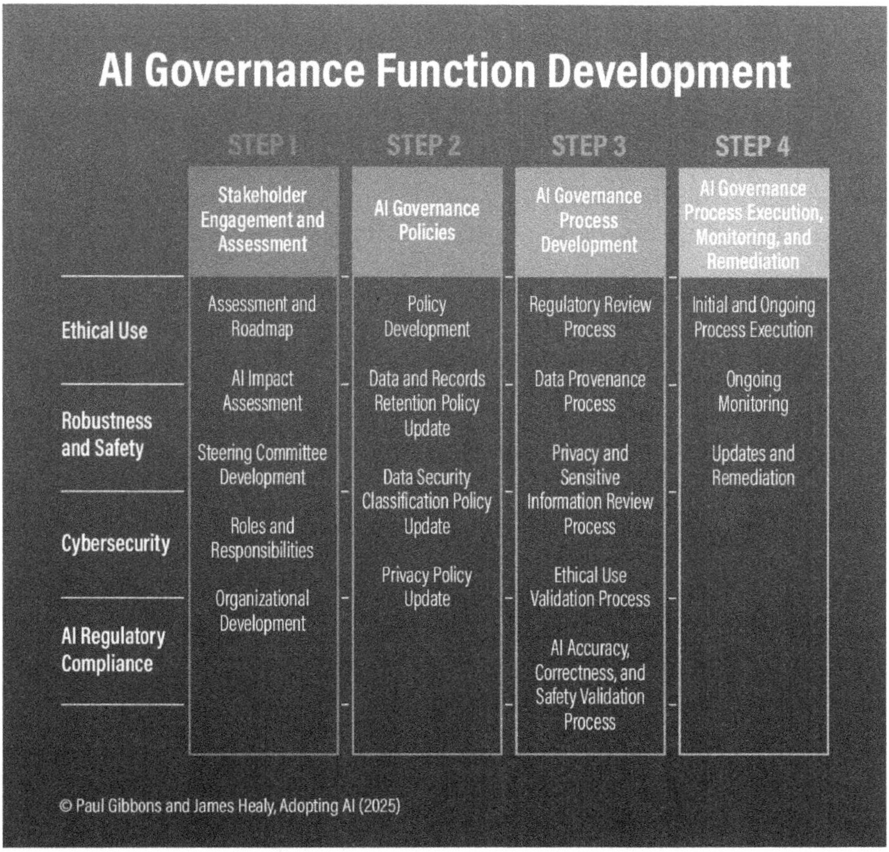

AI Governance Function Development

	STEP 1	STEP 2	STEP 3	STEP 4
	Stakeholder Engagement and Assessment	AI Governance Policies	AI Governance Process Development	AI Governance Process Execution, Monitoring, and Remediation
Ethical Use	Assessment and Roadmap	Policy Development	Regulatory Review Process	Initial and Ongoing Process Execution
Robustness and Safety	AI Impact Assessment	Data and Records Retention Policy Update	Data Provenance Process	Ongoing Monitoring
	Steering Committee Development	Data Security Classification Policy Update	Privacy and Sensitive Information Review Process	Updates and Remediation
Cybersecurity	Roles and Responsibilities	Privacy Policy Update	Ethical Use Validation Process	
AI Regulatory Compliance	Organizational Development		AI Accuracy, Correctness, and Safety Validation Process	

© Paul Gibbons and James Healy, Adopting AI (2025)

FIGURE XI.4: An illustrative AI governance process

Future-proofing Governance

Nothing in any of our business lives has been as complex as AI is at every level, from math and algorithms, through the proliferation of models, through the infrastructure layer, and through the application layer. No technology has had as many vertical and horizontal use cases. No invention will reach more deeply into the employee experience at work. No technology will upend companies, industries and societies as will AI. No new technology has ever transcended the boundaries of mere technology in quite the same way.

Many readers will have suffered through technology releases every several years, who remembers Windows ME, Windows XP, or Windows Vista? With AI, there is a substantive model released every few weeks, not every few years.

Consultants wax lyrical about the pace of change, and how this new "thing" is really "THE THING." Your authors are usually the people saying that the emperor is buck naked, debunking the hypers. But with caution, the hypers might be right this time.

Hence, we worry about the governance gap. Too few leaders understand the technology even superficially, and too few organizations have adapted their governance structures to cater for it. This is understandable. There is a lot of contrarian noise around AI, and experienced leaders have seen fads aplenty come and go.

Wherever companies are on their AI adoption journey, we'd warn that almost none have **more** governance capability than the technology demands. We'd imagine further that there are firms with nearly no specialized AI governance knowledge, praying that their existing risk management, compliance, and strategic governance will provide ample cover for their AI activities. (The AIRS data from the introduction to this chapter places the percentage of firms with AI governance separate from technology governance at 10 percent.)

That seems like an accident waiting to happen.

Organizations that develop robust AI governance will not only avoid ethical pitfalls, but also gain a strategic advantage. They can lead with integrity and foresight and build lasting trust with stakeholders, navigate regulatory landscapes with confidence, and harness AI's transformative potential responsibly.

Future-proofing governance is not just about preventing harm, it is about positioning organizations as stewards of innovation in an AI-driven world, where ethical innovation defines long-term success.

CHAPTER XII

NAVIGATING THE FUTURE WITH A PEOPLE-FIRST APPROACH

Back to the future

As you'll doubtless recall, we started the book with seven pairs of alternative futures outlining dystopian versus utopian scenarios for AI adoption over the next few years. These scenarios aren't predictions, but calls to action. The eleven chapters you've read since are intended to help you tip your organization, and ultimately society as a whole, away from dystopia and toward utopia. (See Figure 0.1 repeated on the next page.)

Parents, citizens, and workers

The grim sci-fi dystopia, Severance, paints a world where personal and working lives are severed by brain surgery. The real world isn't so neatly divided. Because of its vast power, AI will affect us in all our roles, far beyond those as employees or managers, including **at least** our roles as citizens, as individuals with careers to manage, and as parents.

In our view, a parent would be negligent to prohibit a child's natural curiosity toward AI while impairing their future career prospects: it is a powerful learning tool and will become a mandatory skill set (now and certainly in 2030). However, parents must worry about questions such as: Should my children have unlimited access to AI? Should I ban my children from using it until later in their teens, as an increasing number of parents do with social media and smartphones? Should they use it for school? How do I help them navigate the line between research and plagiarism? How can I help them detect bad actors engaging with them for ill intent, but now with powerful AI tools?

AI is not yet a political issue that ranks highly among citizens' concerns, but neither were the internet nor social media for the first dozen years they were available. AI will, in the decade to come, become such an issue, as it should. As citizens, we need to elect representatives who understand the potential risks and potential, and direct policy to maximize the former and minimize the latter.

AI Scenarios: Dystopia or Utopia

DYSTOPIA

Militarized Minds
AI is co-opted for military purposes such as Autonomous Weapons Systems, shifting the focus of scientific research toward conflict and control. This has chemical, biological, radiological, and nuclear risk creating global instability. Researchers face ethical dilemmas as their work is repurposed for surveillance, weapons, and geopolitical competition.

Lopsided Longevity
Breakthroughs in AI-driven healthcare and longevity create treatments unaffordable by all but the super-rich. While the maximum human lifespan is extended, the average remains the same as infectious diseases, obesity, cancer, and tropical diseases remain underfunded and unaddressed.

Breaches and Backlash
Without regulation, AI systems exacerbate bias and create large-scale privacy breaches, causing public backlash, eroding trust, and slowing adoption across industries. Reactive regulatory intervention stifles innovation.

Unemployment and Unrest
Lacking effective labor transition approaches, automation-driven job displacement triggers economic inequality, mass layoffs, and social unrest, creating resistance to AI deployment.

Biosphere Blowout
Energy and water demands of training and running AI models accelerates climate change. Without environmental standards, their unchecked proliferation worsens resource consumption and undermines global sustainability goals.

Automated Absurdity
AI automates tasks and processes that add little to no value: taking minutes of ineffective meetings, summarizing pointless emails, and generating a deluge of mediocre content. Productivity becomes an illusion, with burnout and overwhelm skyrocketing.

Screenbound Solitude
Hyper-personalized AI-driven content fragments shared experiences, creating billions of one-person echo chambers. Increased screen time and diminished real-world interactions lead to a mental health crisis.

© Paul Gibbons and James Healy, Adopting AI (2025)

UTOPIA

Supercharged Science
AI democratizes scientific discovery by providing open-source models and datasets. Collaboration between academia, industry, and government leads to reproducible research and accelerated breakthroughs. AI assists researchers in identifying novel hypotheses, optimizing experiments, and promoting transparency through open science initiatives.

Worldwide Wellness
Preventative care technologies, powered by AI, help combat widespread global health challenges such as infectious diseases, obesity, and cancer. Collaboration between public and private sectors prioritizes broad societal health over profit-only motives. AI doctors provide healthcare access in the world's most remote areas.

Fairness First
AI systems are designed with fairness and inclusivity as core principles. Transparent practices build public trust, leading to widespread, responsible adoption of AI across sectors. Sensible regulation encourages innovation by providing clear ethical guidelines and incentives for responsible AI development.

Evolution, Not Elimination
Governments and industries collaborate on reskilling programs and education reform, preparing the workforce for AI-driven job transformations. Instead of mass unemployment, automation leads to a shift toward higher-value roles, enhancing productivity and quality of life.

Planet Progress
AI technology evolves to prioritize sustainability, with advancements in energy-efficient algorithms and hardware. Environmental standards for AI models are adopted globally, and AI is effectively applied to the greatest sustainability challenges: managing natural resources, combatting climate change, geo-engineering, and supporting conservation.

Creative Catalyst
AI streamlines genuinely valuable tasks, freeing humans from mundane chores and enabling creativity and strategic thinking. Organizations focus on meaningful work, enhancing productivity, job satisfaction, and work-life balance.

Connected Community
AI-enhanced content curation fosters shared experiences and bridges divides. People engage with diverse perspectives while balancing digital interactions with real-world connections, promoting mental well-being and community resilience.

News junkies may recall the farcically ignorant questions asked by members of the US Congress of social media companies in 2018. Such uninformed politicians may default to technophobia, blind techno-optimism, constricting regulations, market-based ethics (whatever that means), or knee-jerk populist reactions, none of which we can afford. On the other hand, nor should we welcome the increasingly cozy relationship between politicians and the broligarchy of Musk, Bezos, Zuckerberg, Altman, and the rest; AI is too powerful to entrust to the titans of tech.

As career-oriented individuals, from university- to middle-aged, workers need to position themselves for an AI future. AI, we have seen, affects every business function and every scientific and engineering endeavor. Even the arts are being transformed. In ten years, **workers without AI competence will likely become dinosaurs**, much less efficient than AI-fluent co-workers. Certain job categories will disappear. The IMF has said that 40% of all jobs in the world could be affected by AI and that it's mostly going to be felt in the white-collar and professional ranks. Our hunch is that this is a gross underestimate.

To remain relevant and resilient, workers must keep the ten-year horizon in view and future-proof their careers. **It is an ethical and economic imperative for workers at every career stage to differentiate themselves from AI, to excel at work that AI can't do and to develop human superpowers that AI can't emulate.**

Power and responsibility

As we bring our journey through AI to a close, we stand at a pivotal moment in human history. Much like Prometheus's gift of fire, artificial intelligence has the power to light the way, offering humanity unprecedented tools for progress and problem-solving. Yet, just as fire can both warm and burn, AI's dual potential demands careful stewardship. It is not enough to adopt AI for AI's sake; we must adopt it with purpose, ethics, and a steadfast commitment to human flourishing.

Our book, *Adopting AI: A People-First Approach*, offers a road-map for organizations, policymakers, and individuals to navigate this complex landscape. We began with a techno-optimistic vision, recognizing AI's transformative potential in every sector, from the sciences to healthcare and education, to finance and creative industries. With our look at the history of AI, we were able to explore issues such as hallucinations and the tantalizing prospect that AI might teach itself in a few days what it takes humans a lifetime to master, just as the chess program AlphaZero did. Our gaze into AI's future, particularly in *AI 2030 and Beyond*, highlighted deep reasoning models and the role of agentic AI, both of which have truly transformative potential.

Throughout this book, we argued that AI adoption should be driven by humanistic values. The chapter on strategy laid out how organizations can align AI initiatives with broader social and ethical goals. Our adaptive adoption approach underscored the importance of understanding the nature of AI, the nature of humans, and the many similarities between the two. We reiterate our conviction that AI is best thought of not as a new technology, but as a new intelligence. In discussing learning, we emphasized the necessity of educating not just technologists but everyone, from frontline workers to C-suite executives, about AI's potential and pitfalls.

We started our treatment of ethics by exploring the ethical issues that follow from the nature of artificial neural networks and large language models: opacity, emergence, and intellectual property issues. These challenges are alien in a world now used to mechanistic computer systems built on symbolic logic, but AI is not such a system. Like the human brain it was originally built to replicate, AI is an impenetrable mystery. The black-box nature of AI reinforces the need for transparency, explainability, and robust governance structures. It also speaks to the broader theme of this book: AI's deployment must serve people, not obscure them.

While organizations are limited in their ability to influence the nature of AI, they have much more agency over other ethical issues;

the balancing act between profit and people is where the rubber meets the road. While AI can drive efficiencies and boost profitability, organizations must resist the temptation to prioritize short-term gains over long-term societal impact. The chapter on AI risks outlined five key dangers from bias and discrimination to security threats, highlighting the need for vigilant risk management. And in examining AI law, we saw how regulatory frameworks are evolving to set guardrails around this powerful technology.

Finally, our introduction to governance brought us full circle. Governance is not just a checkbox exercise for compliance; it is the backbone of ethical AI adoption. Effective governance is about embedding ethics into every decision, ensuring that AI systems are not only effective but also fair, transparent, and aligned with human values. Organizations that approach AI governance with rigor and foresight will not only mitigate risks but also build trust with their stakeholders.

As we look to the future, our message is: AI's destiny is not preordained. We dismissed arguments premised on **technological determinism**, that the technology will make us dumber or smarter, (richer or poorer), more connected or further divided, safer or substantially more imperiled, irrespective of how we choose to use it. **While change is inevitable, whether it represents progress is up to us.**

AI holds up a mirror to humanity. Creating an intelligence to match or surpass our own has been a recurring human dream for hundreds of years. Realizing that dream is undoubtedly a technological triumph, but it is above all a human triumph. Ensuring that future generations regard it as a blessing, rather than a curse, depends on the decisions this generation makes. Those generations will live to see the full flowering of the intelligence transition only if we actively maintain our humanity and ensure that technology remains a tool for human empowerment rather than a force of alienation. Unlike Mary Shelley's Frankenstein, we must remain the creator *and* the master of the strange new entity we have spawned.

We hope that *Adopting AI: A People-First Approach* is a call to action, inviting leaders, innovators, and everyday citizens to engage with AI thoughtfully, ethically, and proactively. The book challenges us to rise to the occasion, embracing AI not just as a technological advancement but as an opportunity to reaffirm our commitment to human dignity, equity, and progress. Like Prometheus, let us bring this gift to humanity with wisdom, humility, and a vision for a brighter future for all.

APPENDIX I

GLOSSARY OF TERMS USED IN ADOPTING AI

Accelerated Workforce

A term coined by Paul Gibbons that describes an ideal, rarely-found culture where organizational learning happens an order of magnitude faster.

Adaptive adoption

Paul and James' framework for AI adoption that replaces change management and emphasizes experimentation, organizational democracy, iterative change, and ethics.

Adversarial machine learning

Techniques designed to manipulate AI models to expose vulnerabilities.

Agent classification

Paul and James' typology categorizing AI agents into seven types: Reflex, Model-Based Reflex, Goal-Based, Utility-Based, Learning, Distributed Intelligence, and Supervisory Agents.

Algorithmic bias

Systematic errors in AI decision-making due to biased data or model assumptions.

Algorithmic transparency

The principle that AI decision-making processes should be understandable and auditable.

Alignment Problem

The challenge of ensuring AI's objectives align with human values.

Anthropic AI

A research company focused on AI safety, creator of Claude.

Artificial General Intelligence (AGI)

A form of AI that can understand, learn, and apply intelligence across a wide range of tasks, much like a human.

Autonomous agents

AI systems that can perform tasks independently, often using reinforcement learning.

Autonomous Decision-making

AI systems that make real-time decisions without human intervention.

Backpropagation

A key algorithm in training neural networks by adjusting weights based on error minimization.

Bayesian networks

Probabilistic models used in AI to represent dependencies among variables.

Bias in AI

The tendency of AI systems to produce prejudiced results due to biased training data or algorithmic design.

Black-box AI

AI systems whose decision-making processes are opaque and difficult to interpret.

Causal inference

Determining cause-and-effect relationships, crucial for AI transparency and fairness.

Cognitive load

The mental effort required to process information, relevant in AI-human interaction design.

Complex Adaptive Systems (CAS)

Systems composed of interacting agents that adapt to their environment, often used to model AI and economic systems.

Computational ethics

The study of ethical implications and decision-making in AI systems.

Constitutional AI

A framework for training AI models based on explicitly defined principles to ensure alignment with ethical norms.

Corporate surveillance

AI-driven monitoring of employees and consumers for data-driven decision-making.

Cybersecurity risks in AI

Threats arising from AI vulnerabilities, including adversarial attacks, model poisoning, and data breaches.

Dario Amodei

CEO of Anthropic, an AI research company focused on AI safety and alignment.

Data augmentation

Techniques used to expand training datasets for AI models, improving robustness.

Deep learning

A type of machine learning using neural networks with multiple layers to analyze complex patterns in large datasets.

Deepfake

Synthetic media generated by AI, often used for realistic but deceptive images or videos.

Differential privacy

A privacy-preserving technique that allows AI to learn from datasets while protecting individual data points.

Digital sovereignty

The concept that nations or organizations should control their AI infrastructure and data governance.

Dysrationalia

The tendency of intelligent individuals to engage in irrational decision-making due to cognitive biases.

EU AI Act

A regulatory framework proposed by the European Union to govern AI development and deployment with risk-based classifications.

Emergence in AI

The phenomenon where complex behaviors arise from simpler AI rules or interactions.

Ethical alignment

Ensuring AI systems align with human values and societal norms.

Explainability

The ability to understand and interpret the decisions made by AI models, particularly in deep learning.

Explainability vs. performance tradeoff

The challenge of balancing AI interpretability with model accuracy and efficiency.

Explainable AI (XAI)

AI systems designed to be interpretable and understandable by humans.

Externalities

Economic side effects of an activity that affect third parties not directly involved in the transaction.

Fairness in AI

Efforts to reduce bias and ensure equitable outcomes in AI decision-making.

Federated cybersecurity

A decentralized security approach where AI models detect threats without exposing private data.

Federated learning

A machine learning approach where models are trained across decentralized devices without sharing raw data.

Generative AI

AI models that generate new content, such as text, images, or code, based on learned patterns.

Geoffrey Hinton (Professor)

Pioneer in neural networks and deep learning, known as the 'godfather of AI'. Winner of 2024 Nobel Prize in Physics.

Hallucination (AI)

When an AI model generates incorrect or misleading information, common in LLMs.

Herbert Simon (Professor)

Renowned behavioral economist and cognitive scientist known for work in decision theory and AI.

Human-Centered AI (HCAI)

An approach that prioritizes human values, user needs, and ethical considerations in AI development.

Human-in-the-Loop (HITL)

A model where AI and human expertise work together to improve outcomes.

Hybrid AI

AI systems that combine symbolic reasoning with deep learning techniques.

Industrial policy

Government strategies aimed at shaping economic development and technological innovation.

Integral change model

Paul Gibbons' systemic change model based on the work of philosopher Ken Wilber that has been adapted for leadership development.

Intelligence Age

A term defining the era in which AI significantly shapes economic, social, and scientific landscapes.

Intelligence explosion

A hypothetical scenario where AI recursively improves itself, leading to rapid advancement.

Intelligence transition

A conceptual shift describing the societal transformation driven by AI, akin to the Industrial Revolution.

Job displacement

The loss of employment due to AI automation and increased efficiency.

Knowledge graphs

Data structures used in AI to represent relationships between entities in a structured format.

Large Language Models (LLMs)

Advanced AI models trained on vast datasets to understand and generate human-like text.

Lethal Autonomous Weapons (LAWs)

AI-powered weapons capable of operating without human intervention.

Machine learning

A subset of AI where algorithms learn from data to improve their performance without being explicitly programmed.

Meta-learning

AI that learns how to learn, improving adaptability across different tasks.

Moral hazard

A situation where one party engages in risky behavior because another party bears the consequences.

Multi-Agent systems

AI systems composed of multiple interacting agents that collaborate or compete.

Narrow AI

AI systems designed to perform specific tasks without possessing general intelligence.

Neural Architecture Search (NAS)

An automated process to optimize neural network structures for AI.

Neuroeconomics

The study of decision-making through the lens of neuroscience, psychology, and economics.

Opacity paradox

The challenge of balancing AI explainability with performance and efficiency.

Opacity in AI

The lack of transparency in AI decision-making, making it difficult to understand or challenge AI-driven outcomes.

Open-source AI

AI models and tools that are freely available for use, modification, and distribution.

People-first AI

Paul and James' philosophy that AI's goals should be directed toward human well-being and prosperity; AI adoption should be adaptive, experimental, and bottom-up; and its ethical constraints be governed by humanistic principles.

Prompt engineering

The process of designing inputs to optimize AI-generated outputs in large language models.

Regulatory capture

When AI companies unduly influence regulators, shaping laws to their advantage.

Regulatory sandboxes

Controlled environments where AI systems can be tested under regulatory oversight before full deployment.

Reinforcement learning

A machine learning technique where an AI agent learns by maximizing rewards. See also RLHF—reinforcement learning with human feedback.

Responsible AI

AI practices focused on fairness, accountability, transparency, and ethics.

Sam Altman

CEO of OpenAI, influential in the development of AI governance and policy.

Self-supervised learning

An approach where AI models learn from unlabeled data without explicit supervision.

Shadow AI

AI usage within an organization without official oversight or governance.

Singularity

A hypothetical point where AI surpasses human intelligence, leading to radical societal changes.

Social credit systems

Coercive AI-driven systems that assign individuals scores based on behavior and compliance.

Stakeholder capitalism

A business model that prioritizes the interests of all stakeholders (employees, society, environment) over pure profit.

Synthetic data

Artificially generated data used to train AI models without relying on real-world data.

Techno-feudalism

A term describing the concentration of AI power among a few dominant corporations, leading to digital monopolies.

Tokenization

The process of breaking text into units (tokens) for processing in NLP models.

Transformer models

A neural network architecture foundational to large language models like GPT.

Turing test

A measure of AI's ability to exhibit human-like intelligence in conversation.

Weak AI vs. Strong AI

Weak AI is task-specific, while strong AI has general intelligence like humans.

APPENDIX II

AI RESOURCES

Books

Philosophical and Foundational Insights

Yuval Noah Harari. *Nexus*

Neil D. Lawrence. *Atomic Human*

Peter Godfrey-Smith. *Other Minds*

George Zarkadakis. *In Our Own Image*

Bostrom, N. (2014). *Superintelligence: Paths, dangers, strategies*. Oxford University Press.

Russell, S. (2019). *Human compatible: Artificial intelligence and the problem of control*. Viking.

Mollick, E. (2024). *Co-Intelligence: Living and Working with AI.*

Tegmark, M. (2017). *Life 3.0: Being human in the age of artificial intelligence*. Knopf.

Social and Ethical Dimensions

Christian, B. (2020). *The alignment problem: Machine learning and human values*. W. W. Norton & Company.

Crawford, K. (2021). *Atlas of AI: Power, politics, and the planetary costs of artificial intelligence*. Yale University Press.

Shneiderman, B. (2022). *Human-centered AI*. Oxford University Press.

Strategic and Business Perspectives

Daugherty, P. R., & Wilson, H. J. (2018). *Human + machine: Reimagining work in the age of AI*. Harvard Business Review Press.

Ganesan, K. (2022). *The business case for AI: A leader's guide to AI strategies, best practices, and real-world applications*. Lioncrest Publishing.

Kissinger, H., Schmidt, E., & Huttenlocher, D. (2021). *The age of AI: And our human future*. Little, Brown and Company.

Lee, K.-F., & Qiufan, C. (2021). *AI 2041: Ten visions for our future*. Currency.

Periodicals, newsletters, and blogs

The Economist. (n.d.). The go-to read for cabinet and board members worldwide, superb on business, geopolitics, and technology. https://www.economist.com

DeepLearning AI newsletter. Andrew Ng's excellent perspectives on AI development. https://www.deeplearning.ai/the-batch/

MIT Technology Review. Excellent application and societal writing on AI. https://www.technologyreview.com

The Gradient. Technical and social analysis of AI. https://thegradient.pub

Berkeley AI Research Blog. (BAIR) An accessible, general-audience platform for BAIR researchers. https://bair.berkeley.edu/blog/

Marcus on AI. Thoughtful informed skepticism. *https://garymarcus.substack.com/*

Academic Journals—Ethics, AI and Society

AI and Ethics. *Springer.* Retrieved from https://link.springer.com/journal/43681

Journal of Responsible Technology. *Elsevier.* Retrieved from https://www.journals.elsevier.com/journal-of-responsible-technology

Ethics and Information Technology. *Springer.* Retrieved from https://link.springer.com/journal/10676

Technology in Society. *Elsevier.* Retrieved from https://www.journals.elsevier.com/technology-in-society

AI & Society. *Springer.* Retrieved from https://link.springer.com/journal/146

Must-read papers

"Attention Is All You Need" (2017) - Introduced the transformer architecture that revolutionized NLP and underpins most modern large language models.

"Machines of Loving Grace" (2024) Dario Amodei—Strikes a lovely balance between optimism and caution. Beautifully argued.

"On the Dangers of Stochastic Parrots: Can Language Models Be Too Big?" (2021) — Bender, Gebru et al.'s critical examination of large language models' limitations and social impacts.

"Deep Learning" (2015) — LeCun, Bengio, and Hinton's seminal Nature paper surveying deep learning fundamentals.

"Sparks of Artificial General Intelligence: Early experiments with GPT-4" (2023) - Microsoft researchers documented surprising capabilities of GPT-4 across multiple domains, suggesting large language models might exhibit precursors to general intelligence rather than merely specialized performance.

"Language Models are Few-Shot Learners" (2020) - Introduced GPT-3 and demonstrated emergent abilities in large language models.

"Artificial Intelligence as Augmented Intelligence: An Ethical Framework" (2019) - Important ethical perspective on AI development as enhancement rather than replacement of human capabilities.

Podcasts

West, S. (2013–present). *Philosophize This!* https://philosophizethis.org Explores all of philosophy but with excellent episodes on consciousness and AI.

Fridman, L. (2018–present). *Lex Fridman Podcast* https://lexfridman.com/podcast/ Probably the most prestigious guest list in the technology, politics, philosophy, and science podcast world, and superb questions from its host, but with a heavy far-right political slant.

MIT Technology Review. *In Machines We Trust.* https://www.technologyreview.com The far-reaching impact of AI on our lives.

Harris, S. *Making Sense* https://www.samharris.org/podcasts/making-sense-episodes Incredible breadth and intellectual courage.

The Gradient. (Producers). *The Gradient Podcast.* https://thegradient.pub In-depth interviews with researchers and builders.

Healy, J. (Host). *The B-word Podcast.* https://open.spotify.com/show/5DnTKoyOk9HWSDuCnHN8a5 Behavioral science, organizational change, and the human condition.

Gibbons, P. (Host). (2015–2018.). *Think Bigger Think Better* https://open.spotify.com/show/6ivBVNmBu841cSgJwx503u Philosophy, science, economics.

People to follow

James Wilson—Accenture Managing Director and serial author including Human + Machine.

Andreas Horn—Head of AIOps at IBM and insightful commentator.

Ethan Mollick—Author of Co-intelligence, Wharton Professor, with superb applied AI insights.

Cassie Kozyrkov—Chief Decision Scientist at Google, a leading voice in decision intelligence and data science.

Kate Crawford—Atlas of AI author and AI ethics researcher.

Global and not-for-profit organizations to follow

World Economic Forum (WEF) The WEF engages global leaders in discussions about AI governance, ethics, and the impact of emerging technologies on the workforce.

Partnership on AI (PAI) PAI is a multi-stakeholder organization fostering research and discussion on AI's ethical and societal implications, emphasizing fairness, accountability, and transparency.

Future of Life Institute (FLI): FLI works on ensuring that AI and other emerging technologies benefit humanity, with a particular focus on mitigating existential risks.

AI Now Institute A research institute based at NYU, AI Now examines the social implications of AI, focusing on bias, labor impacts, and AI's integration into public systems.

The Alan Turing Institute: As the UK's national institute for data science and AI, the Alan Turing Institute leads research on AI ethics, governance, and the social impact of AI technologies.

Future of Humanity Institute (FHI): FHI conducts interdisciplinary research on global catastrophic risks, the ethics of AI, and strategies for ensuring the long-term survival and flourishing of humanity.

AI RESOURCES

APPENDIX III

ACKNOWLEDGEMENTS

We are both grateful to Rehouven Libine, Sania Khan, Felix Baumeister, Theresa Moulton, Eloise Seidelin , Josie Maurel, Neil Harrison, Michael Bungay Stanier, Jess Tayel, Jon Z Bentley, Robert Meza, Samuel Salzer, Peter Slattery, Guy Champniss.

Paul's two sons, Conor and Luca, have shown just how fast AI adoption can be for the curious minded. His almost-90 dad, Willie, was unwaveringly supportive, both when Paul was a young scientist and throughout. His brother John, as a cybersecurity expert, vetted that entire section. His sisters, Geri and Fiona, chipped in advice along the way. He'd also like to thank dear friends Don Mayer, David Bennett, Dan Blum, Allen Broome, Alan Arnett, Adam Gold, Clayton Hamm, Tricia Kennedy, and Frank Smits for their decades of support.

James would like to thank Carla and Noah for their patience and support for his at times obsessive writing behaviours. Similarly, thanks to his Mum and Dad for their patience and support for his at times obsessive reading behaviours as a child, without which none of this would have happened. He's hugely grateful to all those at Deloitte who helped shape his ideas about AI through conversation, challenge, and co-creation: Gavin Bergheim, Kim Bracke, Dan Cogan, John Eikland, Ben Fish, Veronica Holmes, Max Ivanovic, Ian MacGregor, Steve Paola, Caitlyn Roberts, Ash Samson, Emma Stuart, Shyanne Strydom, Simon Thuijs, Kirsten Watson, Marcel Wilson, and Yves Van Durme. He thoroughly enjoyed and was greatly influenced by conversations for the B-Word podcast episodes that touched on AI and adjacent topics: thanks to Joe Devln, Robin Dunbar, Patrick Fagan, Rory Sutherland, Dale Whelehan, and George Zarkadakis.

APPENDIX IV

HOW WE USED AI

On the one hand, not using AI to write a non-fiction book would border on negligence. On the other, it is a terrible writer. Here, in the spirit of disclosure, and perhaps as a resource for other writers, here are some of the ways we used AI.

- ☞ Providing strategy and tactics for securing non-English language publishers.

- ☞ Fact-checking entire chapters.

- ☞ Producing a glossary of terms.

- ☞ Doing research on dozens of topics, from Greek mythology, to neural networks, to CRISPR, to economics.

- ☞ Finding up-to-the-minute research material.

- ☞ Generating an outline from forty pages of notes.

- ☞ Copyediting and proofreading.

- ☞ Making a table from a page of text.

- ☞ Researching a list of AI resources and formatting in AP style.

- ☞ Summarizing 150-page technical reports like those from Stanford and the WEF. Providing helpful, though often insufficiently critical feedback on draft chapters.